Creating a Greater Whole

Best Practices and Advances in Program Management Series

Series Editor
Ginger Levin

Creating a Greater Whole

A Project Manager's Guide to Becoming a Leader

Susan G. Schwartz, PMP

CRC Press
Taylor & Francis Group
Boca Raton London New York

CRC Press is an imprint of the
Taylor & Francis Group, an **Informa** business

AN AUERBACH BOOK

CRC Press
Taylor & Francis Group
6000 Broken Sound Parkway NW, Suite 300
Boca Raton, FL 33487-2742

First issued in paperback 2021

© 2018 by Taylor & Francis Group, LLC
CRC Press is an imprint of Taylor & Francis Group, an Informa business

No claim to original U.S. Government works

Printed on acid-free paper

ISBN-13: 978-1-138-06405-8 (hbk)
ISBN-13: 978-1-03-209594-3 (pbk)

Library of Congress Cataloging-in-Publication Data

Names: Schwartz, Susan, 1959- author.
Title: Creating a greater whole : a project manager's guide to becoming a
leader / Susan Schwartz.
Description: New York : CRC Press, [2018] | Series: Best practices and
advances in program management series
Identifiers: LCCN 2017038408| ISBN 9781138064058 (hb : alk. paper) | ISBN
9781315160641 (e)
Subjects: LCSH: Leadership. | Project management.
Classification: LCC HD57.7 .S39 2018 | DDC 658.4/092--dc23
LC record available at https://lccn.loc.gov/2017038408

Visit the Taylor & Francis Web site at
http://www.taylorandfrancis.com

and the CRC Press Web site at
http://www.crcpress.com

To Dave, my true north

Contents

Preface

Leadership is not a solo sport. It is about people and how they engage with each other—not the individual leader or his/her associated ego. Leaders recognize and polish the strengths of the individuals who make up a community. Whether your community is a professional work environment or a volunteer organization, it is the power of collaborative capabilities that enables people to make sense out of chaos, overcome challenges, and achieve their shared goals.

Strong leaders are big picture thinkers who have a clear vision that inspires and motivates people to engage and actively pursue an initiative. They are able to identify people's strengths, communicate passion, and project confidence to build a team that works together to make success happen. They have innovative problem-solving and clear communication skills. Their uncanny ability to correctly identify and focus on core issues, communicate in a powerful way, and neutralize the potential conflict seems so easy. It is not!

As experienced leaders know, the hardest part is maintaining calm so that people can logically work through the challenges at hand. It is only through trial, error, and mistakes that leaders have honed these skills. The best leaders have all made mistakes; the key is they do not consider a mistake to be a failure because they learn from their experiences and take the acquired knowledge forward. They are lifelong learners who encourage the people around them to join them on the discovery journey. Leaders ask questions to seek diverse perspectives, and when necessary, redefine the rules.

As I was making my own journey from manager to leader, I did not understand how people made leadership look so easy. I felt as if I was making it up as I went along. I studied the actions of my managers and mentors to learn how they made it all look so simple. I started reading every book on leadership that I could find. What I discovered was that leadership is all about continuous learning because the only environmental condition leaders can depend on is change. And yes, sometimes leaders do make it up as they go along because there is no precedent—the situation truly has not happened before.

I realized my *hero* leaders were keen observers who asked questions and engaged their teams as active problem-solvers. I learned that leadership is not acquired by osmosis or inoculation. Leadership acumen is acquired through practice, mistakes, adaptation, and repetition. My purpose in writing this book is to offer project managers and general managers a different perspective and some tools they can use to more easily navigate their own leadership journeys.

This book is meant to be a guide that provides a different viewpoint to help managers who are transitioning from tactical oversight positions and want to take on strategic, vision-building responsibilities. Although the content was written with a specific thought flow in mind, it is very possible to read sections in any order and still gain from the benefits of new perspective and reflection.

Chapter 1 begins by exploring different traits attributed to leaders. Continuing forward, Chapters 2 and 3 discuss how to work through the leadership basics of aligning your vision and communicating your message to enable people to understand everyone's roles so that they are able to begin working collaboratively. Chapter 4 addresses how to handle people with different communications and working styles. The conversation is expanded in Chapter 5, which describes how to select the best motivational method based on the needs of an individual and the specific situation. Chapters 7 and 8 explore the various stressors of uncertainty, risk, and conflict and some tools that leaders can deploy to turn potentially explosive situations into constructive conversations. Chapter 9 focuses on the human element of leaders. People look toward leaders when the going gets tough. This chapter focuses on how you, as a leader, should take time to focus on yourself and your professional development. Leadership is not about following well-defined rules. Leadership is about helping people navigate toward the unknown. It is not necessarily about taking Robert Frost's proverbial path less traveled. It is about forging new paths by leveraging the strengths of your team to build the necessary roads and bridges for those people coming next. At the end of the day, leadership is defined by the people (aka community) who make the journey together. Leaders are the people who maintain alignment among the various organizational parts as the community members work together to achieve their shared mission and vision.

Creating A Greater Whole is the culmination of many travels. I do not believe this book is an end. I like to think of it as a milestone for reflection and a launch point for future expeditions. There were many people with whom I worked along the way who are too numerous to mention in this small space. To all those who traveled with me, thank you for sharing the journey and helping me navigate across the bumpy moments. Together, we were able to make some good things happen, which none of us could have done on our own.

This book is one of those projects that could not have happened without the encouragement, advice, and counsel of trusted friends. Thanks to Terry Tuley, Susan Kudla Finn, Sandra Bernhardt, CC Clark, Rizwan Shah, Cynthia Huheey, Beth Spriggs, Matt Lowy, Ellen Savel, Ken Cohn, Victoria Guido, Jessica Burbach, Naomi Otterness, Deb Peer, Bryan Calabro, and John Pan who all were generous extending their time, perspective, and feedback.

Finally, thanks to my family who were always there at the end of a long day to remind me what is truly important. My husband, Dave, was my first line editor for this book. Thank you for always being direct and kind. My children Carl and Emily, and son-in-law, Aaron, who were my strongest cheerleaders when I asked them if they thought I could write a book. A special shout out to Emily who after an early read asked, "What happened to my mother's voice?" The initial drafts were too formal. She reminded me to be authentic and true to the ideas of leadership and honesty I shared with her and her brother growing up.

Author

Susan G. Schwartz, The River Birch Group, has led a variety of technical and nontechnical teams successfully through organizational change and brought projects to completion on time and under budget. Over the course of her career, she developed unique methods to motivate cross-functional, global teams by combining process review, crucial conversation techniques, and knowledge-sharing strategies.

She has facilitated public project management and leadership courses and seminars for clients such as George Mason University Learning Solutions, the Project Management Institute, The American Society of Quality, Association of Talent Development, and the Society for Human Resources.

Ms. Schwartz received an MS in Organization Development and Knowledge Management from George Mason University School of Public Policy, Fairfax, Virginia and a BA in Economics from Goucher College, Baltimore, Maryland.

Chapter 1

What Is Leadership?

Leaders are defined by …
Trust enables leaders to …
Leaders aren't necessarily born …

Surprise, You Are a Leader!

Effective leadership traits are easy to describe but difficult to measure. People say they know strong leaders when they see them, but are not able to define just what it is that makes a leader successful. Not everyone who takes on leadership responsibilities follows a preestablished career path. Sometimes leaders arise from the most surprising circumstances. A person can be very happy being a member of a team. Then, something unexpected happens and without too much thought, he or she realizes a need and steps forward to fill the void. They may have minimal experience and may be uncertain of their capabilities; but, voilà! Suddenly, they are a leader.

Have you ever found yourself suddenly in a leadership position? My first opportunity to lead a project occurred the first time I shadowed an experienced systems analyst for a large telephone system implementation. I was 23 years old, and the woman I was shadowing had 10 years of experience. Her pregnancy became difficult, she was placed on bed rest, and suddenly I was in charge. The senior technicians who oversaw the equipment installation were grizzled curmudgeons who looked me up and down and announced their granddaughters were older than me. What could I do? I did not pretend to be an expert. I worked hard to perform my role as the customer and engineering liaison and asked them a great many questions. Those two men took pride in their crusty natures. Once I had earned their respect, they were very kind and helpful as we worked together to resolve the problems that always happen during a systems implementation. They taught me the value of hard work, humility, integrity, and trust. Of course, the twice weekly box of doughnuts that I bought on my way to the job site may also have helped to build the comradery that evolved into a trusting relationship.

Leadership: "I Know It When I See It"

Leadership development can be especially difficult in technical fields. More often than not, once people achieve technical expertise, there are few options open to them for promotion or advancement besides management. These high performing people come to this new role with minimal experience overseeing operational activities or supervising junior staff, and even less training. Yet, they are expected to intuitively know how to build teams, motivate people, manage conflict, negotiate resources, and prioritize risks.

Discovering unexpected knowledge requirements for leaders is a common situation across nearly all fields. A highly skilled physician told me about the learning curve that she encountered with the 'business' of healthcare activities when she took over as the medical director for the practice. She was expert on human physiology topics but was at a loss when she had to weigh in on a marketing decision. Everyone cannot know everything about everything. I asked her how she was closing her knowledge gap. Her answer was akin to one meeting at a time.

The Project Management Institute (PMI) defines leadership as "the ability to guide the project team while achieving project objectives and balancing project constraints" (Project Management Institute 2013, p. 17). The Human Resource Management portion of the PMI industry standard guidelines identifies the interpersonal skills needed to analyze a situation and then leverage the strengths of team members to achieve the project goal. These skills are categorized by the areas of *leadership*, *influencing*, and *effective decision-making*. The PMI guidelines state the importance of clearly articulating a vision, the need for listening skills, the pursuit of various perspectives, and the value of a decision-making process (Project Management Institute 2013, pp. 283–284). Everyone agrees that leaders need these skills; but, how does a project manager learn and gain competency in these areas? It would be so much easier if we could attain these skills simply by reading a book or going to a few classes. Instead, we must figure it out for ourselves through the successes and failures that we experience along the way.

So how does one learn to be a leader when leadership traits are hard to discern, difficult to measure, and require unfamiliar skill sets? The little spoken secret regarding leadership success is that most people have leadership potential within them. Some people have easy access to these internal leadership wellsprings and others may need to work a little harder to bring them to the surface. The method for both types of people is the same: observe, reflect, apply, and repeat.

Observation is the key for learning cause and effect. How does someone in a leadership position react to a challenge? If you are fortunate and are able to work under a strong, positive leader, pay attention to how the leader handles a situation and speaks with people involved. If you are working with a difficult manager/leader, you can still take away some very strong lessons. Observe how this leader handles a situation and speaks with people. Then, file away these examples of how you will NOT manage/lead when placed in a similar situation. Reflection is the process of analyzing the positive and negative consequences that can result from the different actions you could pursue before deciding how to handle a challenging situation.

Reflective Exercises

Often the best means of honing your leadership skills is to reflect back on your experiences to recognize similar situations and consider the actions you would like to model moving forward and those actions you want to be sure not to repeat. Your reflection time may be limited to sitting in a traffic jam, waiting in a checkout line at the grocery store, or other situations you might consider time-wasting. Just the opposite—these are valuable moments that you can use to reflect on any variety of complexities.

The *Consider and Deliberate* exercises throughout this book were developed to provide you suggested questions that are geared to help you reflect on different leadership concepts and scenarios. You may insert questions of your own that are more applicable to your specific situation or experience. There is space for you to write your reflective comments, or you may prefer to keep a separate journal. The purpose of the workbook format is to help you avoid future fatal errors by reflecting on situations leaders experience and using these experiences to act more effectively when confronted by a challenging situation. It is simply a start to your reflective process. I encourage you to return to these reflections six months or a year from now to see how your experiences and opinions might have changed.

Leadership Basics

When people are asked to identify traits that strong leaders possess, they may list 15–20 off the top of their heads. These often include: courage, vision, excellent communication, integrity, honesty, expertise, and charisma. Strong leaders possess many of those skills; however, it is impractical to try to acquire them all at once. If something appears too complicated, people might give up without even trying.

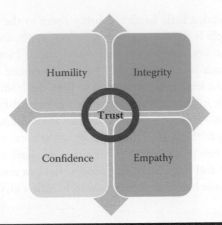

Figure 1.1   Diamond Leadership Model.

Like a golf swing, it is better to focus on honing the fundamental skills before trying to play a challenging course.

The Diamond Leadership Model illustrated by Figure 1.1 identifies five core competencies of extraordinary leadership. Just as every diamond is different, with varying levels of cut, clarity, color, and carat, leaders are not stamped from a single standard of excellence. When I have queried people who tell me that they know leadership when they see it; they share anecdotes that represent the intangible facets: trust, confidence, humility, integrity, and empathy.

Trust is an essential leadership attribute that interweaves the other four leadership competencies. At the core, people follow leaders because they trust them to perform with the best interest of the people who work with them at heart. If a leader does not engender trust, the other competencies do not matter.

Of the four primary facets, *humility* is one of the most important, in that true leaders realize everyone, even him/herself, possesses strengths and weaknesses such that an individual cannot usually succeed alone. *Empathy* is the ability of a leader to put him/herself in another's position. Empathetic leaders are able to view a challenging situation from a variety of perspectives and consider the impact an action may have on all associated parties. *Integrity* is defined as performing to expectations in that you follow through with what you have promised to do. The final facet of *confidence* is a matter of competence as opposed to expertise. The leader may not be a subject matter expert; however, she/he knows who and how to get things done. When several things go wrong simultaneously, people want their leader to maintain calm and tell them what to do. Even if the leader does not have a clue, she/he will orchestrate everyone on the team to begin the diagnostic and resolution process. The most successful leaders not only appear confident but also are able to encourage others to become confident.

Trust

Brené Brown explores the courage that it takes to lead groups within her book *Daring Greatly*. She tells many wonderful stories; however, my favorite metaphor that describes the trust exhibited by a fully functional team is the act of "turning toward each other" as opposed to "turning on each other" (Brown 2012, p. 101). Bad things happen. They just do. When trouble happens within dysfunctional teams, the team members pull away from each other and point fingers to put blame on others. When functional teams hit a snag, they exercise their trust bond and pull together to work toward a constructive resolution.

Humility, integrity, empathy, and confidence may be difficult characteristics to measure; however, economists and neuroscientists have targeted the impact of trust within work environments as the focus for several research projects. Stephen M.R. Covey recognized a direct relationship between trust and the speed and cost required to accomplish a designated work effort. This relationship is demonstrated by a simple formula:

$$\text{Trust} = \frac{\text{Speed}}{\text{Cost}}$$

When trust is low within an organization, it takes more time and costs more money to get things done. When trust is high within an organization, actions are achieved more quickly and at lower cost.

He extends his economic driver metaphor to incorporate trust taxes and trust dividends. Covey levies an 80% trust tax on organizations where trust is nonexistent. These organizations are described as having a toxic culture. People within these organizations actively sabotage other people's work efforts, and management focuses on punishing negative activities as opposed to encouraging positive outcomes. As trust improves within an organization, the tax decreases to zero. Covey awards trust dividends to organizations that exhibit trust-based environments. He recognizes a world class trust environment as one that promotes collaborative innovation, transparent communications, employee engagement, and systems alignment (Covey 2006, pp. 13–25). Why would any leader not want to aspire to creating a high-trust organization?

Paul Zak, a neuroeconomist, combines economics, psychology, and neuroscience to fully understand how humans make decisions. His research focuses on how neuroscience research can be applied in a business environment to build high-performing, high-trust groups. His findings reflect similar results to those discovered by Covey and his team. He found that people who work within high-trust organizations exhibit more energy and are more productive than people who work within lower trust organizations. Not surprisingly, the people who worked within organizations that maintained a culture of trust reported lower turnover rates and a positive work–life balance. Comparing a broad spectrum of U.S. companies, Zak's team discovered that the average trust metric was 70% out of a possible 100%. The two lowest scoring elements tracked by the researchers were recognizing excellence and sharing information (Zak 2017, pp. 87–90). The positive takeaway is that leaders can improve the trust within their workplaces with minimal effort and budget by using transparent communication techniques and leveraging their innate skills for humility, integrity, empathy, and confidence building.

Humility

Humility is one of the first traits listed when people describe the leaders they most admire. Proverbs and time-honored adages advocate that challenges make leaders stronger, better, courageous, and humble. One of my favorite sayings is, "that which doesn't kill us makes us stronger." However, I have learned that some traits which we might consider a weakness cannot always be eradicated via hard work; sometimes, we are who we are. And that is why humility is such an important trait for leaders to acknowledge. Leadership is not about perfection. Leadership is about helping other people navigate difficult situations. You need to be honest with yourself and others when you need to reach out to someone else who might have stronger skills in a specific area.

John C. Maxwell, who the *New York Times* recognizes as a leading expert in the field of leadership, explains humility with a reference from the pastor Rick Warren, "Humility is not denying your strengths. Humility is being honest about your weaknesses" (Maxwell 2014, p. 33). A strong historical example where a leader was able to be successful by being honest about his weaknesses was described by Emmet Murphy in his book *The Genius of Sitting Bull*. He identified "Strategic Humility" as one of Chief Sitting Bull's strongest leadership traits that enabled him to unite the independent Sioux tribes to defeat George Armstrong Custer at Little Big Horn.

In 1868, the United States Government signed a treaty with the Sioux Nation that guaranteed them ownership of the Black Hills territory. Today, this territory is known as South Dakota; however, this land had belonged to the Sioux Nation for several centuries before the 1868 treaty was signed. Troubles began in 1874 when gold was discovered in the Black Hills. Suddenly, prospectors began to ravage Sioux territory in search of gold. The federal government tried to renegotiate the treaty and offered to purchase the lands. The Sioux tribal chiefs refused to consider any offer. The tribal lands were not for sale. In 1876, the United States War Department authorized operations to commence against the *hostile* Sioux.

Up until this point the tribal chiefs of the Sioux nation were independent governance units and maintained relatively peaceful alliances. In sizing up the strength of General Custer and his troops, Sitting Bull evaluated his strengths and his weaknesses. He also studied the strengths and weaknesses of the other four Sioux tribal chiefs: (1) Crazy Horse, (2) Gall, (3) Four Horns, and (4) Red Cloud. The U.S. troops had many more soldiers and much more fire power that the Sioux possessed; however, Sitting Bull was convinced that, by coordinating their leadership strengths and not fighting separately, they could provide an insurmountable force to defeat Custer and his Bluecoat forces. Chief Crazy Horse brought innovative creativity, which fostered brilliant guerilla warfare tactics to the leadership cohort. Chief Gall's contribution was well-organized, in-depth planning strategies. Chief Four Horns' contribution was his ability to make connections and build relationships that provided competitive intelligence which could be used against the enemy. Chief Red Cloud, the most senior of the chiefs, brought rational discussion founded on years of experience and factual analysis to assure all angles had been considered. Sitting Bull exhibited his unique collaborative skills to build a ring of trust as he convinced the four Sioux chiefs to join together to execute his strategy, enabling each of his potential rivals to head up different groups based on their particular strength without losing control or respect (Murphy 1996, pp. 51–88).

Figure 1.2 illustrates how the unique traits of each of these Sioux leaders who comprised the multifaceted leadership combined to create a whole strategy. These same skill areas continue to be applied to situations today where people with varying strengths contribute their energies to create a greater whole than they could imagine achieving as an individual.

When trust is present and people are able to relax their guard and share their honest, off-the-cuff thoughts, collaborative brainstorming and networking activities will drive the creation of innovative solutions. These solutions have a greater probability for successful execution because the managers involved are able to be honest with each other when analyzing challenges and developing short- and long-term plans.

Leaders need to be aware of their strengths, their motivations, and their priorities as well as their weaknesses to understand and motivate other people. The first *Consider and Deliberate* exercise offers you an opportunity to take a moment to reflect on the various elements that comprise your unique personality.

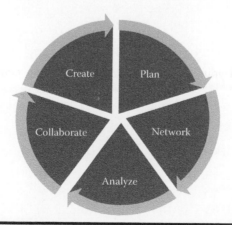

Figure 1.2 Leadership ring of trust.

CONSIDER AND DELIBERATE

What three words would you use to describe yourself?

Describe a scenario that illustrates each of the selected descriptive words.

List two areas that you consider weaknesses for yourself. Do you accept each one as part of your organic self? Given the right motivation, what can you do to improve these skill areas?

What skill would your colleagues say you most often contribute to group efforts?

What brings you joy?

How do you channel that joy?

It is relatively easy to recognize the strengths and weaknesses of colleagues and friends. Self-evaluation is never easy. There is no right or wrong answer. This reflection is a tool that you can use to help you understand the personal strengths from which your leadership abilities can develop. The final two questions are important to ask yourself at the end of each day. What brought you joy today? Did you achieve a specific goal? Perhaps you shared in an accomplishment or witnessed a random act of kindness? Did someone impact your day or did you impact someone else's day? Was it as simple as hugging a loved one? How did you share that positive experience with others around you? People who recognize the small moments are that much more astute when it comes to addressing the larger ones.

Integrity

As one of the foundation leadership facets, it is important not to confuse integrity with behaving ethically. Several years ago, I saw a sign in a shop window in Vancouver, Canada that read, "Al Capone may not have been ethical, but he had integrity." I did a double take. On reflection, I realized Al Capone, who was an infamous gangster during the 1920s, may not have followed the prescribed laws and behavioral mores of the time; however, you could count on him to follow through with any action he promised to dole out. If you welched on a debt and he told you he would break your legs, you knew you had better have some crutches on hand. How many of your colleagues are law-abiding citizens who consistently have excuses why they cannot meet deadlines or never follow through on promised commitments? These people certainly are ethical, but they have no integrity as they cannot be counted on to do what they say they will do; when they say they will do it.

In addition to following through on commitments, integrity is demonstrated by leaders who perform according to the same rules they set for the people with whom they work. Informally, people recognize a leader's integrity by saying someone walks the talk. They do not make empty speeches demanding people to do more with less and then go back to their ivory tower offices with catered lunches. Emmett Murphy provides an example of how not to lead when he describes General George Custer's demand that his soldiers march in full-buttoned uniform carrying heavy ammunition in the hot summer heat. Any deviance from this strict, but pointless, dress code was severely punished. Meanwhile, Custer rode among his troops in a lighter, summer-weight uniform with no equipment and changed to a fresh horse when the one he was riding became tired. He treated his horse with more concern than he showed his reporting troops. In fact, Custer outlined two sets of rules: (1) one he demanded his troops follow and (2) another for himself. In the fury of the battle with Sitting Bull and the Sioux warriors, many soldiers realizing Custer was safe and well behind the line of battle broke ranks to save themselves. They felt no loyalty to General Custer because Custer had never shown any care or loyalty for them (Murphy 1996, pp. 74–76).

Empathy

Some people consider empathy to be a sign of weakness. They believe if you do not demand toughness at all times or if you ask for another opinion, the people reporting to you will consider you weak and not worthy of respect. This is a total misconception. Empathy is not the act of over-sympathizing. Empathy is recognizing how others are impacted by an action. It is about paying attention to the consequences of your decisions. It does not diminish a leader's authority or control.

In fact, empathy is all about paying attention. There is a very powerful line from *The Death of a Salesman* when Willy Loman, who has spent his whole career working for a clothing company, is unceremoniously retired. In his anguish, he screams, "Attention must be paid! Attention must be paid!" He wanted a small piece of respect. He wanted an acknowledgment of his humanity. Empathy is paying attention to the people involved and the impact of the consequences from your decisions—both positive and negative.

When evaluating a new process or procedure, empathetic leaders try to consider how the people who contribute to the process and the people who receive the outputs of the procedure will be impacted by the change. When they reach out to people who can provide a broader view of the problem to help develop the solution, these leaders guarantee the solution will incorporate the full organizational perspective to provide a stronger solution than if they had charged forward with a single viewpoint. Strong relationships with long-term allies who stay with you during the tough times are another benefit that empathetic leaders can reap by practicing inclusion among colleagues and partners.

The preparation for the Battle of the Little Big Horn demonstrates the difference between the leadership styles of General Custer and Chief Sitting Bull. Custer did not believe there was a need to prepare or train the troops following him into battle. Communications and explanations were minimal, if any. His focus was centered on himself and his personal needs. Sitting Bull, on the other hand, proactively sought out the viewpoints of the various Sioux tribes. The methods he used to motivate them to join together were relatively simple. First, he reached out to establish communications, and then he spoke of his appreciation for the strengths of each party, and then discussed the commonality the groups shared (Murphy 1996, pp. 76–78). His three part communication strategy helped him demonstrate his desire to build a relationship and share the goals they could work toward together. Empathy helped Chief Sitting Bull to seek out a broader focus that enabled him to develop a winning leadership strategy.

Confidence

Some people have a natural swagger whether they know what they are talking about or not. These charismatic people only talk the talk as opposed to walking the talk. Shakespeare's line from Macbeth, "...full of sound and fury, signifying nothing" is an apt description for some people. They will step forward to grab the leadership role; however, they are not able to generate confidence among the people assigned to work with them.

Leaders know how to get things done. They are aware they cannot make everything happen on their own. A confident leader has experience and knowledge that is recognized by the team members, colleagues, and stakeholders to guide everyone's efforts to achieve the designated goal. Their confidence comes from developing an ability to recognize people's talents to create a synchronized effort in which the whole is significantly greater than the sum of the individual parts.

Strong leaders spend much more time asking questions and listening intently than telling people what to do. John C. Maxwell sums this concept up when he says, "Any leader who asks the right questions of the right people has the potential to discover great ideas" (Maxwell 2014, p. 13). Confidence emerges when people know there are members on their team who can answer the questions for which they do not have the answers. The confidence of knowing that colleagues will support each other during uncertain times is a result of leaders melding the facets of humility, integrity, empathy, and confidence to create the trust necessary to develop and maintain healthy teams.

CONSIDER AND DELIBERATE

For each of the five facets that comprise the Diamond Leadership Model, briefly describe a scenario that reflects a situation in which you were a participant.

Trust:

Humility:

Integrity:

Empathy:

Confidence:

Were your examples all drawn from the same situation?

If not, what similarities do you recognize that are common to each of the scenarios?

For positive scenarios, how can you assure the experience is repeatable?

If you considered a scenario to have a negative outcome, what can you do to assure a positive situation occurs the next time a similar situation is encountered.

Different situations will require leaders to draw on different skill sets. In fact, once you become comfortable and have honed one area of leadership expertise, the unexpected will happen and you will need to hone a new set of skills to help your group successfully address a challenge that may never before been addressed. Focusing on creating and maintaining a culture of trust, even in the most chaotic of situations, will provide an environment that encourages all of your reporting team members, colleagues, and stakeholders to synchronize their best efforts and successfully achieve the mission.

Instinctive Leadership

Some people are born with a dominant leadership gene. They are the people who were class president and kickball team captains. Other people are 'pegged' as team players. Good team players are hard to find and valued people. However, be careful about putting yourself in a box. It is a great experience for perennial leaders to be a team player once in a while to gain a different perspective and experience a different energy. And sometimes, great team players will find themselves in a leadership position. Given support, these people will often bring a fresh and energizing perspective to the group. The following two scenarios (one real life and one fictional) describe situations in which an 'average Joe' who never considered becoming a leader found that he was leading fairly complex work efforts.

Jay Coakley and Ellie's Hats

Jay Coakley never planned to become a leader. He had spent his career as a middle school gym teacher. He is a humble man who was the kind of gym teacher we all wish we had. He got bored in retirement and found a position as a part-time elementary school gym team teacher. He met a five-year-old kindergartener named Ellie who was undergoing chemotherapy. She always wore a hat and a big smile. Jay went home and told his wife he would like to give her a hat. They ordered a cute knitted cap from a website. A few of their friends heard about the hat they ordered for Ellie and wanted to contribute more hats. Then, a knitting circle at his church heard about Ellie and wanted to donate hats.

Suddenly, he had more hats than Ellie could possibly wear. He contacted Ellie's mother who helped put him in touch with INOVA Fairfax hospital. His contact suggested the annual winter holiday party would be a good place to distribute the hats. As word continued to spread, knitting circles from as far away as New Jersey and South Africa were sending hats. Requests began to come in from all over the country for hats for pediatric cancer patients. A local restaurant helped him sponsor a fundraiser to pay the shipping costs. Sports teams began donating caps. Jay found himself managing an international team of volunteers. He then received several large checks from people who wanted to help but did not knit or crochet. He spoke with his contact at INOVA Fairfax and began a program that provided welcome bags and iPads to pediatric oncology patients. He realized he had done something phenomenal when the nonprofit he started, which he named Ellie's Hats, presented the INOVA pediatric oncology program with a $10,000 check.

Jay also helped lobby the Virginia legislature to create a Childhood Cancer Awareness license plate. He received a request from a very ill little boy who wanted his parents to order a pediatric cancer license plate. When his parents discovered there was none, they appealed to Jay. He knew nothing of the required process, but he knew people who he thought might know. He called them and asked for their help. Suddenly, Jay was a lobbyist. Jay is quick to attribute his success to the many people who shared their expertise with him when he did not know how to accomplish a mission task. In only 18 months, Jay Coakley who had never considered becoming a manager became a leader heading up a 501c3 international nonprofit group that now lobbies and raises funds to support children who are fighting cancer. To think it all started from Jay purchasing a single hat.

Leadership Lessons from the *Lego*® *Movie*

If you are not familiar with the *Lego Movie*, I will quickly summarize the plot. The movie takes place in an urban environment ruled by a dictator-like ruler, President Business, and his evil alter ego, Lord Business. The protagonist, Emmett, is a construction worker who follows the same schedule every day. He is very happy to follow the rules set forth by President Business. A resident wise man prophesizes that a 'yellow face' who is referred to as 'The Special' will rise up to save the world. Emmett is accidentally identified as the prophesized savior. There are several side stories, which include an independent female Master Builder named Wild Style who creates the most amazing vehicles out of Legos. She introduces Emmett to the other Master Builders who want to overthrow President/Lord Business. After a false start, Emmett becomes an accidental leader of the team because he can follow directions and adapt the plan when the team runs into a snag. He works with the Master Builders to deploy the plan that successfully overthrows Lord Business. President Business is 'freed' and asks to join Emmett's successful team.

The four primary project leadership lessons I gleaned from the movie that can be used by both new and experienced leaders are as follows:

1. Success is attainable when you believe in your team and yourself.
2. Just because "that is the way we've always done it" does not mean that is the only way. People should not be afraid to stand up and suggest an innovative idea.
3. Creative people may be very smart; however, success often requires many skill sets. Subject matter experts need to develop professional skills to guide and motivate project teams toward success.
4. A plan that leverages available resources and talent is the essential project foundation.

Plans should not be set in stone. Maintaining a flexible, can-do attitude helps you to evaluate and adapt the original plan to changing situations and keep your focus on the goal.

CONSIDER AND DELIBERATE

How are Jay Coakley and Emmett similar?

Describe a situation when you have watched a single individual (perhaps yourself) address a small need that grew into something larger than you anticipated.

Which of the four project leadership lessons would you select first for your leadership tool kit? Describe a situation where it might have helped to resolve a challenging situation.

What other tool(s) should be included in Emmett's leadership tool kit?

The real-life Jay and fictional Emmett are very practical people. They did not go in search of a spotlight. They began their leadership journeys by proactively seeing a need and looking to how they could close the gap. They became leaders because they were able to involve people who had the needed expertise for different tasks. Next, they coordinated the skills and knowledge of the team members to create a whole that could not have been built by the group members had they been working individually.

Summary

What is Leadership?

- Leaders are people who are able to build teams, motivate people, manage conflict, negotiate resources, and prioritize tasks.
- The Diamond Leadership Model identifies the five primary facets of extraordinary leadership: (1) humility, (2) integrity, (3) empathy, (4) confidence, and (5) trust.
- A culture of trust is created by leaders who are able to meld the facets of humility, integrity, empathy, and confidence.
- Humility is the ability to be honest with yourselves and others regarding your strengths and weaknesses.
- Integrity demonstrates a leader's ability to follow through on their commitments.
- Empathy helps a leader recognize how all affected people will be impacted by a decision.
- Confidence is a competency skill that enables leaders to recognize people's talents to create a synchronized effort to achieve a shared goal.

Chapter 2

Aligning Teams and Steering the Journey

Leaders align action and vision by ...
Boundaries help leaders focus ...
Setting priorities enables leaders to ...

Leadership Requires More than a Title

Keeping teams focused and on-track can be difficult during normal operating circumstances. When situations occur that are outside your team's control, the immediate goals can seem nearly impossible. As people's stress levels rise, their attentions may diverge as they begin to look to escape from the situation at hand. It becomes the leader's primary responsibility to calm their team's worries and help them to refocus and align their energies with the mission.

Chris Hadfield overcame many challenges to become the first Canadian astronaut and then a Commander of the International Space Station. He describes a leadership lesson he learned during one of the NASA training exercises that put his team into survival situation in which they were dropped on top of a mesa, "good leadership means leading the way not hectoring other people to do things your way." He discovered being the loudest voice did not assure that people would listen. He learned that leading his peers required a different skillset. He needed to know how to guide people through the stresses of a difficult situation. He also realized that empathy and a sense of humor can be very helpful leadership tools during difficult situations (Hadfield 2015).

Although major business changes may not entail the same type of life and death circumstances faced by astronauts, in the moment, the crisis being handled will significantly affect everyone involved and can feel like 'life or death.' Leaders need to acknowledge the challenge, mitigate as many concerns as they can, and focus the teams on where they need to go.

In June 2013, Colonel Scott A. Jackson learned that he would take command of the third Infantry Division's second Armored Brigade Combat Team that had just been designated as one of ten army brigades to be downsized as part of the budget reduction mandated by the Budget Control Act of 2011. The U.S. military had not gone through this type of exercise in many years, certainly not in the memory of these soldiers. The Spartans, as this unit was known, returned from a tour in Afghanistan to learn the Brigade was to be *downsized* and assumed this meant they were being *fired* from the army. Jackson knew he needed a plan that addressed the basic concerns for the 3800 soldiers who reported to him which included the following question: did they have a job and if so, would their working units be transferred intact, or separated?

The low morale of the Spartan Brigade and the families in attendance at the Change of Command ceremony was evident as Colonel Jackson stepped forward to accept the responsibility of the brigade command. He began his command acceptance speech by addressing the core concerns of his new team and their families, "The first thing that's most important for you to know is that you and your families are going to be taken care of. Nobody gets fired in the Army, so every one of you who choose to stay in the Army will have a home." He assured the soldiers that most of them would stay within their units and be absorbed into the other third Infantry Division's brigade. Once he had assured the soldiers their basic needs would be addressed, Colonel Jackson finished his speech by telling them their immediate mission was to train as hard as they could to maintain and improve their readiness so that when they transferred from his command, they would be reporting to their new assignments as the best trained soldiers in the U.S. Army. His acceptance of command speech ended with the Spartan team shouting the Spartan motto, "Send Me," at the top of their voices to demonstrate their commitment to achieve the mission their Colonel outlined. In less than three minutes, Colonel Jackson reenergized the brigade, confirmed their value to the army, and aligned their focus toward the immediate actions that would help them get to the next step of their journey.

Colonel Jackson then began the intense work of developing a plan that would assure smooth personnel transfer and the divestment of 52,000 pieces of equipment. He attacked this mission with the same energy with which he pursued every other mission demonstrating his five core personal values: (1) people matter; (2) we only succeed when we work as a team; (3) maintain discipline which means approaching routine tasks with a common and high standard; (4) integrity—do the right thing because it is the right thing to do; and (5) balance your multiple priorities of work, health, and family. He led by example and used continuous communications as his primary leadership tool to help the Brigade maintain focus over the next 17 months. He provided his team information as he received it. When the decommissioning date was set, Jackson began to carry index cards with each soldier's name and assignment as it was set. Any soldier could stop the Colonel to ask after an assignment. If the solder voiced a concern that Jackson could not answer, he made a commitment to find the answer.

When Colonel Jackson officially deactivated the second Brigade on January 19, 2015, 68% of the team had been absorbed into the other Brigades at Fort Stewart, Georgia. The working units (approx. 2580) people remained intact. Most of the rest of the brigade retired or made a career transfer. A very small number of people (2%–4%) were required to transfer outside the original unit because they held highly specialized skill sets which could not be used at Fort Stewart.

The entire process happened over 17 month. The smooth transition can be attributed to the nine month of precise planning that preceded the eight-month deactivation process. The overall goal which stayed constant was to maintain a state of readiness and atmosphere of trust among the cohort of soldiers. Colonel Jackson attributes the smooth and successful transition to the strategic vision provided by the ability of his senior officers to manage changing priorities and the expert execution performed by the junior officers. He viewed his Brigade Commander role as that of the chief communicator, who assures everyone has the information they needed to perform at their best (Scott A. Jackson, personal interview, May 26, 2017).

Managing and Leading

Whether you find yourself in the private sector, public sector, or military, managing and leading groups of people is incredibly hard work. Whether your role is that of a supervisor, a manager, an executive, or a strong team player, functional leadership does not mean you get to order people around. Being *in charge* means being where the proverbial buck stops. It means you are accountable for the actions of the people around you. It means you need to figure out how to motivate your team or orga-nizational group to come together to deliver goods and services within expected scope, budget, and schedule. It means you need to continually observe the actions of the internal and external stakeholders around you to be ready to initiate action to maintain organizational balance and achieve established mission goals.

Very often the leader is the hardest working person of the team. Cartoons show-ing executives with their feet up on their desk while everyone else scurries around does not represent reality. But what does it take for managers and leaders to be suc-cessful? Are they the same job function? What does it take to be a good manager? What additional skill sets are required by a strong leader?

During 2016, the Project Management Institute (PMI) put a stake in the ground by issuing professional education parameters that organized project-management-related skills among three areas: technical, business management, and leadership. Technical topics are fairly easy to identify, but how do you organize topics between business management and leadership? Some people distinguish between management and leadership topics based on focus and priorities of the person performing the work.

By using this lens, managers are the people who keep the wheels of defined possibilities oiled and working smoothly, whereas leaders find people with the right skills to make what seems like an improbable vision possible. Overlay people who are trying to lead across matrix teams have all of the responsibilities without specific authority over resources a complex, three-dimensional mosaic based on requirements, which can change from moment to moment.

Reflect back to people who you would try to emulate as a leader. It could be a historical figure, a relative, a project leader, or significant team member. What skills or traits would you try to model?

CONSIDER AND DELIBERATE

What makes a strong manager?

What makes a great leader?

Name four people who you consider to be great leaders.

List management and leadership skills you feel you need to develop for your professional development. Identify the top two skills in each category you would like to prioritize to include in your professional development plan.

There is no right or wrong answer to this reflection. When I facilitate this exercise during a class, people often list manager traits that involve technical expertise, budgeting skills, resource management, clear communications, and respectful behaviors. However, when they brainstorm a list naming great leaders and the reasons why they selected these people, they describe abilities that instill confidence and convince people they have what it takes to achieve a difficult challenge. The management skill sets and knowledge requirements vary from class to class based on the industry and experience level of the course participants. The nontangible leadership skills are the same in every class. The key take away for me from my anecdotal research is that leadership is about people. Technology, tools, and procedures will change, but people's desire for respect, appreciation, and accomplishment never goes out of style.

Ground Zero for Leaders

Ralph Waldo Emerson is credited with the phrase, "Life is a journey, not a destination." Extending Mr. Emerson's metaphor, leadership is a meandering journey that starts at a leader's personal ground zero. Just as you and your team become comfortable on a path toward your goal, some type of change will occur, and you will need to determine an alternate means to get there. A leader needs to be confident of his/her starting point or 'ground zero' and have a fairly good idea of how they relate to the mission's end goal. Some people call this point a person's *true north*. When challenges or environmental shifts alter the initially planned roadmap, the leader's primary responsibility is to make sure their actions and those of the group remain aligned and contribute to the end goal.

Once you are aware of your 'ground zero' and have an idea of your immediate path, you can begin to help the people working with you to find their ground zero. The following reflective exercise is designed to help you tackle one of the toughest challenges for a leader which is creating a plan for yourself.

CONSIDER AND DELIBERATE

Describe your professional/personal status today.

Describe where you want to be professionally/personally a year from now.

What is your plan to reach your destination(s)?

No matter how complex and multifaceted an organization grows, strong leaders are aware of the direction for their own journeys. They set time aside for professional development and are open to new learning opportunities.

Aligning Communities for Success

Once you have identified your ground zero, the next step is to learn about the people who will be traveling with you on your journey. The PMI defines anyone who is associated with a project as a stakeholder. I challenge you to recognize these individual stakeholders as the community of people who will help you achieve your designated goal. It is the combined efforts of all stakeholders that will enable successful project completion. If each stakeholder only focuses on his or her individual responsibilities, success cannot always be assured.

Some people use the terms village or tribe to describe a group of peoples' interrelationship. Whether people reference community, village, or tribe, the constructive energies developed from these groups engender the trust and confidence that will achieve the shared goal. Nadia Bolz-Weber, a twenty-first-century theologian, describes this concept best when she said, "I don't think faith is given in sufficient quantity to individuals necessarily. I think it's given in sufficient quantity to community." Leaders need to make sure that the people who they are leading understand the ground zero for the group and that each individual realizes their own personal ground zero within the organization community and how their role(s) interconnect with the other community members.

Simon Sinek describes the protection people receive when they are part of a coordinated community as the Circle of Safety. He attributes the comradery that develops among tightly connected people as the power that enables collaborative problem solving. Each person is so attuned to their interlocking activities that when one notices something is awry, they signal everyone there is a problem and combine their resources to resolve the issue. If an outside danger, such as someone intent on undermining leadership's efforts tries to enter the Circle of Safety, aligned groups will shrug off the negative endeavor. When groups are not aligned and only pay attention to their specific tasks, the danger may not be recognized until the damage incurred is past the point of repair (Sinek 2014).

When the unexpected happens, people look toward leaders to maintain focus and keep people on track. In the midst of uncertainty and doubt, leaders are the ones who maintain calm and keep people pointed toward the designated destination. They prioritize the important factors and redirect energies that are being wasted on trivial matters. Who were the people in your life, other than your parents, whose actions or words made you feel confident in the face uncertainty? They might have been a coach, a teacher, a community leader, a manager, or a client. Reflect back on wise words they shared with you that you still repeat to yourself today. How do you model those behaviors? As a counterpoint, for a very brief moment, reflect back on those actions, you found so abhorrent that you promised you would never repeat them when you were in a leadership position. How do you prevent yourself from modeling poor behaviors?

CONSIDER AND DELIBERATE

In your career, has someone stood out as a very strong leader, or a very weak leader? List their names in the space below.

Briefly describe an instance of positive or poor leadership behavior.

In hindsight, what about the positive leadership examples did you find reassuring? What was missing from the poor leadership behaviors?

Describe a situation in which you demonstrated strong leadership skills based on lessons you took from one of the above examples.

A common denominator among strong leadership examples is the leader's ability to encourage behaviors that will help the group to focus on a common purpose or mission. Weaker leader examples often describe someone who does not take responsibility for decisions, blames others, or stirs up conflict among group members. Great leaders understand that group success is really about a community of people who come together to make something happen. A leader's primary role is to assure that everyone's individual effort align with the organizational mission and purpose.

Alignment and Leadership Global Positioning Systems

Just as Global Positioning Systems (GPS) systems help drivers navigate unfamiliar routes, leaders provide a similar function to align and integrate the efforts of their teams as they navigate unfamiliar territory. When people are spread across distances and organizations, it is easy for them to get sidetracked and adjust priorities. The leader's role is to maintain steady communications and to know at what time they should intervene to let people know when and why they need to *make a legal U-turn*.

Dr. Henry Cloud discusses the importance of leaders being able to set boundaries. He defines boundaries as "what you create and what you allow" (Cloud 2013, p. 14). The concept of boundaries can apply to micromanagement, accountability, or organizational culture. Boundaries work well for all team members because they set an agenda for communications. The designated interconnection points enable people to identify who needs to be included in pertinent discussions.

Boundaries become tools that enable leaders to say "No" so they can focus on the current priorities. As managers, many of us believed we had to say "Yes" to all but the most outlandish requests. We would just work longer and harder to get what needed to be done. As you transition to leadership functions, your workload only gets more complex. It is a matter of survival that before agreeing to a request, you take note on how the request aligns with the mission designated for you and your team. Boundaries will help you hone your neural GPS-like functions so that you can focus your efforts on three basic leadership tasks (Cloud 2013, p. 29):

■ Pay attention to what is important.
■ Inhibit what is not important.
■ Keep team current and on track.

Aligning Teams

The first step toward integrating the efforts of a group or team is for a leader to establish a common purpose and assure that each person knows how they fit into the total effort. I refer to this step as the "why are we here?" and "why do we care?"

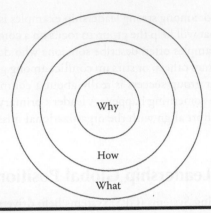

Figure 2.1 Sinek's Golden Circle.

phase. By aligning each person's efforts with the overall mission, it is much easier for everyone to realize what is important and ignore what is not in order to stay on course. If this process is skipped over, the leader will spend many wasted hours *recalculating the route* to get the group back on track.

Simon Sinek uses the construct of a Golden Circle, in his book *Start with Why—How Great Leaders Inspire Everyone to Take Action*, to describe how leaders can provide a balanced effort of motivation and execution techniques by identifying the *Why, How, and What* of actions and decisions. Figure 2.1 illustrates Sinek's Golden Circle theory. *Why* is located at the inner core of the Golden Circle and offers people clarity of vision in order for them to understand the purpose and background information that drives the designated action. The outer circle designates the *What* which is the desired outcome or product. *How* is the middle circle that connects the *Why* and *What* layers of Sinek's Golden Circle. This layer describes the methods and disciplines needed to take the concept from a vision to reality (Sinek 2009, pp. 65–67).

Have you ever been given a task that did not make sense or was not part of a normal procedure? When you questioned the rationale or correctness of the direction you heard, did the person issuing the command tell you it was not your job to ask why? How well did you perform that task? Developing strong communication skills and avoiding these types of situations is the basis for Sinek's concept of the Golden Circle and the balance between *What* and *Why*. People need to understand how they fit within the greater vision of the organization. No, it is not necessary to share the dressing down you might have received by your boss because things are delayed through no fault of your own. But, what is the story? Why is the task at hand important? Once people grasp the vision, the *How* layer provides the standards, guidelines, and procedures that are used to create the final work product.

Leaders must be confident that people have the skills, understanding, and motivation to perform the tasks and activities required to achieve the shared mission.

If any of these attributes are missing, it is the leader's responsibility to close the gaps. If skill sets are missing, they arrange for training or coaching support to assure the people on the team achieve the necessary competency.

When people do not understand their role in achieving the end goal, take a hard look at your communication techniques and the information that you share. If motivation is lacking, initiate a discussion with the affected parties. The solution may be very simple, or it might be a symptom of a deep-rooted problem. Do not let these nagging issues fester among your team members. The situation will only explode at a most inopportune moment.

The Organizational Clarity Model

The Organizational Clarity model illustrated in Figure 2.2 is based on the premise that organizations are made up of different entities that may or may not overlap. The cone represents the enterprise-wide 'umbrella.' The various entity spheres combine to create the multiple aspects of the organization. The expanded sphere within the second cone highlights the three leadership tiers or layers that exist within each of the organizational entities. The inner tier focuses on the priorities for the leader of the group. The second tier addresses the priorities of the people who are within the leader's immediate working group. The third tier encompasses the strategic priorities shared by all of the groups that comprise the organizational entity. The graphic can be adapted as organizational layers increase or decrease.

The basic priorities of each tier, which the leader needs to consider when aligning the vision and purpose for the group, are outlined in Figure 2.3. The inner-tier traits focus on how a leader views his/her internal self and how he/she reacts to the external

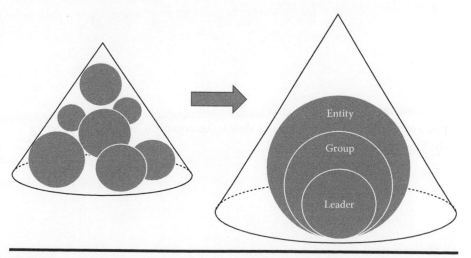

Figure 2.2   The Organizational Clarity Model.

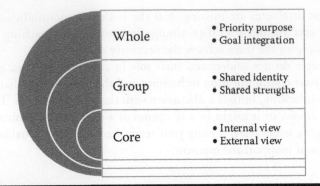

Whole	• Priority purpose • Goal integration
Group	• Shared identity • Shared strengths
Core	• Internal view • External view

Figure 2.3 Aligning group vision and purpose.

views proffered by working group members. The middle tier demonstrates the leader's role in leveraging the various strengths and skill sets of the immediate working team to develop a shared group identity and share each other's strengths. At the outer tier, leaders work together to integrate objectives and align the priorities of the groups assembled within the entity organization. This level of the model is where the group's role within the organization is defined as it integrates with the overall mission.

CONSIDER AND DELIBERATE

Select an organization you work within as a part of your professional or personal life.

Place yourself at the inner core and sketch the organizational structure using the Organizational Clarity model.

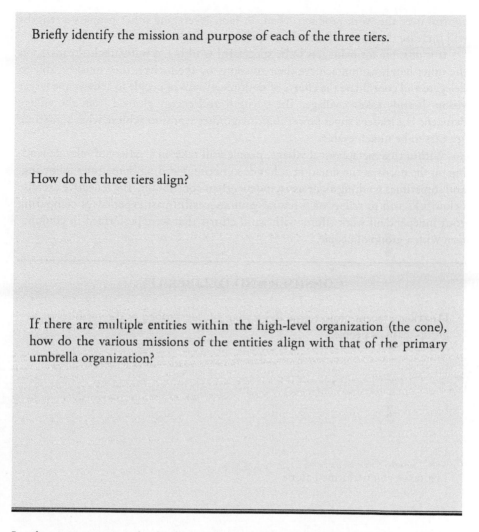

Briefly identify the mission and purpose of each of the three tiers.

How do the three tiers align?

If there are multiple entities within the high-level organization (the cone), how do the various missions of the entities align with that of the primary umbrella organization?

Leaders must constantly think at a variety of levels incorporating potential consequences. What appears to be a simple situation can become very complicated as people from multiple entities begin to work together. Essentially, a leader becomes an organizational GPS able to identify obstacles and quickly reroute the original plan to achieve the designated goal as expeditiously as possible.

Leadership and Delegation

Some managers find it difficult to delegate. Either they do not have enough staff to perform the task, or they believe they can do it faster themselves. And then, there are those managers who are afraid to delegate authority, because they will lose control over the work product; when, in fact, leveraging other people's strengths will increase the power of the effort being extended.

It is possible for managers to be successful working as individuals. Leaders, on the other hand, cannot achieve their missions by themselves; they must be able to delegate and coordinate the efforts of multiple groups of people to achieve the larger vision. It truly takes a village. The strength and energy gleaned from the village denizens is a leader's super power that invigorates teams to achieve what sometime appears to be unachievable.

Within this metaphorical village, people will take on a variety of roles depending on the needs of the situation at hand: sometimes leading, sometimes following, and sometimes working alone as an independent contributor. The reflective exercise below asks you to reflect on a few of your successful past experiences comparing your independent work efforts with work efforts that were performed in conjunction with a group of people.

CONSIDER AND DELIBERATE

Describe a recent project or activity that you performed as the member of a group/team of people.

List tasks you performed alone.

List tasks you performed as part of a group.

How did these two types of tasks differ?

How would you describe your efforts with regard to the two different types of tasks?

What skills did you use working as a group that you did not need when working alone?

Working with people toward a common goal can be energizing or incredibly frustrating. Did you prefer working on the individual tasks or did you prefer the group-based efforts? When you think back to your experiences working with groups of people, which end of the spectrum did you find yourself? If you had an energizing experience, did everyone understand their role and how the various work efforts connected together? If you had a frustrating experience, was everyone working in different directions?

Givers and Takers

Often strong leaders are described as people who always take the time to stop and listen no matter how busy they might be. They are described as being generous and open to suggestions. In his book, *Give and Take*, Adam Grant describes why helping others can drive a person's success. Grant expands on the basic concept of networking and project stakeholder relationships by dividing people into three categories: (1) Takers, (2) Matchers, and (3) Givers. Takers are described as people who rarely recognize the needs of others. Matchers specifically pay attention to *quid pro quo* (I will do this for you, and you will do this for me). Givers are described as people who are more concerned about helping others without regard to themselves.

Each of these three categories of people is able to use their strengths to be successful leaders within different environments. Takers have proven to gain the most positive results when the results are based on a finite win-or-lose outcome. Givers are considered to be stronger leaders when they work with longer term, complex situations. Some explain the basis for these strengths as the Takers' ability to focus in on the essential needs of the moment, whereas Givers hone their talents to build relationships and establish long-term trust among people.

Based on the presumption that Givers by their nature prefer to help others, no one anticipated the results of Grant's research that recognized the top performing people to be consistently identified as Givers. At the opposite extreme, the lowest performing people also were identified as Givers. Takers and Matchers were ranked within the middle of the performance bands. What is the difference among various Givers that would cause such a significant performance gap? Grant's research identified this differentiating factor that so widely separates Givers between the top and bottom performance categories as the amount of self-interest a person holds for their personal achievement illustrated by Figure 2.4. The bottom performing Givers are *Selfless* individuals, who care so much about helping others succeed, place minimal importance on their own rankings. The higher performing Givers are identified as *Otherish*. These individuals enjoy helping other people achieve success; however, they are able to maintain a balance that allows them to achieve their individual performance goals.

Leaders who are Otherish Givers value the time they share with their team members and associated stakeholders, because they recognize this time as an investment. The relationships they build participating in giving activities provide them

High	Low
Otherish	*Selfless*

Figure 2.4 Self-interest.

future value above and beyond the immediate act of giving. Moreover, they understand that listening to other people's perspectives is both a means to expand their viewpoint and a way to show respect for the individual's contributions toward the team goal (Grant 2013, pp. 6–158).

Summary

Leadership and Community Alignment
- Group success happens when people come together with the purpose of attaining a shared vision.
- Strong leaders help people recognize their roles, relate to each other, and integrate efforts.
- The primary role of a group leader is to maintain what is important, minimize what is not, and keep people on track.
- Leaders cannot achieve their missions by themselves; they must be able to delegate and coordinate the efforts of multiple groups of people to achieve the vision at hand.

Chapter 3

The Power of Collaboration and Conversation

Groups get things done by ...
Collaborative communities are ...
Conversation is powerful when ...

Coming Together for Strength

I grew up enjoying Charles Schultz's comic strip, *Peanuts*. Charlie Brown and Snoopy were the main characters; however, as the fourth of five children, I particularly related to Linus who was bossy Lucy's younger brother. One storyline that has stayed with me had Linus asking why he should do what his sister told him to do. Lucy told him, "There are five reasons." She then counted her fingers and curled them into a fist. Linus immediately fell into line and did what she wanted him to do. Bullying aside, the concept of a fist is a great metaphor for individuals coming together to be something stronger than they can be individually. Individually, we are similar to the fingers on a hand. We have unique fingerprints and flexibility; however, when fingers merge into a fist, a more powerful entity is created.

The title of the current book, *Creating a Greater Whole*, was inspired by Aristotle's adage "the whole is greater than the sum of its parts" that describes the synergy that happens when various entities (or people) merge to create positive energies. Strong leaders help collaborative teams identify their unique capabilities and integrate the efforts of all the participants into a unified force to achieve the shared vision.

Whether you are working on a neighborhood event or a complex global project, the foundation for collaborative success is determined by how people value each other's contributions. A positive outcome is contingent on the respect each group member has for everyone's contributions. Often the reason a project fails is because group members do not respect each other, which means they have no foundation on which to build trust. They spend more energy looking over shoulders or erecting silos than they would if they had taken time to learn about the other members and the strengths they bring to the team.

Adam Grant discusses the complexities of developing organizations, which encourage creativity and innovative thinking in his book, *Originals.* His research identified skills, potential, and cultural fit as the three most important factors that organizations consider during the hiring process. In fact, many of the participating organizations identified cultural fit as the most important element they considered when making hiring decisions (Grant 2016, pp. 180–181). Grant cautions leaders who are trying to build innovative organizations to not hire on cultural fit, because they may find themselves in a situation of groupthink as many of their new hires will bring similar thinking into the organization. Instead, he recommends leaders who are trying to build cultures of originality should focus hiring decisions on the cultural contribution a job candidate will bring to the organization. Hiring managers should consider what traits and skill sets are needed to broaden the perspective and decision-making diversity of their group (Grant 2016, p. 251).

This is the power of a collaborative community. The best of everyone's abilities are pooled together to create something much bigger than could be created by people working alone. A single superstar cannot generate the same level of innovation and value as a group of individuals working together.

The power of collaboration is the combination of people working through a constructive process to resolve a challenge. Although good things happen when people collaborate, leaders must be cognizant of the investment of time and effort.

When urgency is the priority criteria, leaders may not be able to defer to a collaborative process and will need to shoulder the full responsibility of the decision. But when time and knowledgeable people are available to work through the issues, the resolution will incorporate a variety of viewpoints and the people taking part in the process will a have a stake in the process and own the solution.

CONSIDER AND DELIBERATE

Describe a time when you had to resolve a difficult situation by yourself.

What was your first thought?

What was the outcome? How was it achieved?

Describe a time when you had to resolve a difficult situation as part of a group.

What was your first thought?

What was the reaction of the group?

What was the outcome? How was it achieved?

What different types of accomplishment do you associate with each of these situations?

Solving difficult situations as part of a collaborative group usually is easier than working through a problem alone. Yes, there can be conflict and extended discussions; however, what seems to some as 'time wasters' are often a prelude to innovative solutions.

Culture of Responsibility

When people come together as a group to develop a solution, there will be discussions about expectations, rights, and responsibilities as illustrated in Figure 3.1. Experienced managers are adept at defining scope and requirements that define expectations. Rights are defined by regulatory and legislative bodies. Responsibilities are more difficult to specifically identify within groups undergoing change because of the variety of work cultures and experiences people bring with them. An organization's culture of responsibility often is defined using terms such as control, accountability, blame, independence, and decision-making.

Rigid, tightly monitored organizations are described as cultures that focus on blame and control. Collaborative teams are most successful when they are able to work within a culture that encourages individual accountability and enables

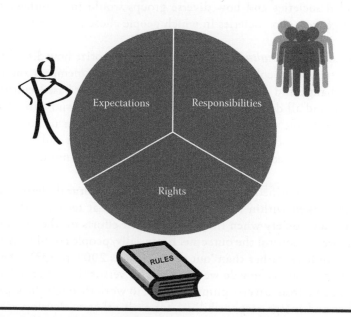

Figure 3.1   Culture of responsibility.

decentralized decision-making. One measure for a successful leader is how well newly merged teams are able to work together to adapt intrinsic cultures and meet the needs of the evolving community.

Building a Responsible Community

Jonathan Sacks writes about the culture of responsibility in his book, *The Home We Build Together: Recreating Society*. He began writing the book as a reaction to the global events post-9/11. At that time, he was the Chief Rabbi of the United Hebrew Congregants of Britain and had seen much of the forward-moving, multicultural strides Britain had made begin to erode. He was concerned by what he saw as societal breakdowns and asked the question, "How can we rebuild the structures of our life together in pursuit of the common good?" He launched his journey by defining society as "the realm in which *all* of us is more important than *any* of us ... It is where we come together to achieve collectively what none of us can do alone" (Sacks 2007, pp. 4–5).

As he explored society-building through the centuries, Sacks referenced historical, political, and theological perspectives. Responsibility and freedom of choice were defining themes across all three disciplines. Sacks initially discussed responsibility as a matter of choice by an individual and how these small decisions impacted the larger society. He then expanded his exploration to global societies and how diverse groups would interconnect to begin building large, complex societies in which people chose to value the dignity of difference.

He coined the term *culture of responsibility* to categorize how different groups defined individual and collective responsibility to group members and outsiders to manage change and create a social order. Individually based cultures build personal silos and all decisions focus on the self. Familial- and tribal-based cultures develop specific rules and responsibilities that pertain to how people are treated within those communities and a separate set of rules and responsibilities for the treatment of people who come from outside the community (Sacks 2007, pp. 132–133).

His analysis identified the most successful multicultural change agents as people who worked within responsibility cultures that recognized they could only build a new society when they focused their efforts on the areas in which they had power to control the outcome. Essentially, people could be responsible for "change 'in here' rather than 'out there'" (Sacks 2007, p. 133). The twenty-first-century question is how do we adapt the 'us versus them mentality' so that people who come from diverse groups are able to work through the separateness of the other and merge responsibilities for each other to develop a collective force that will foster constructive change that will help to achieve the community's goals.

CONSIDER AND DELIBERATE

Describe a situation (educational or professional) in which you were part of the majority and someone from another type of culture was assigned to join your group.

How was the other person welcomed into your group?

Did anyone behave differently toward this person?

How would you describe the culture of responsibility demonstrated by the group?

Describe a time when you found yourself in a new cultural situation in which you were the other.

Was your first thought to escape, to blend in, or to hold onto the difference?

How did the people around you treat your *otherness*?

Did anyone offer you advice or social cues?

How long did it take for you to stop feeling like the other and feel like a member of the group?

When people build societies, all members of the founding group do not always share the same culture of responsibility. Prior experiences and traditions can make a difference. People observe that young children find it easier to 'blend in' than adults. What is it that enables young children to be more accepting of the diversity of others than many adults? Perhaps it is because they have minimal life experience and no preconceptions which allow them to be more trusting, and, therefore, they are able to identify common ground more quickly. Adults do not have easy access to playgrounds and kickball fields, however, they can find common ground through conversation.

Choices and Conversation

People are described as accountable when they acknowledge their role in a decision and shoulder the consequences of any associated actions. Mistakes happen. The mark of a strong leader is what they do when these mistakes happen. Sometimes the people who must rectify the mistake had nothing to do with the cause. When this happens, how do people within your organization react? Do they say, "Not my fault, not my job to fix?" Or does your team come together to resolve the challenge with a broader focus? Taking responsibility to fix your mistake or to help someone else resolve a challenge is a choice.

Leadership and responsibility are defined by choices that people make. When I teach leadership classes, I use two examples about choice and responsibility that I learned from my children. The first involves my daughter when she was three years old. One morning, as I was trying to get us breakfast and out of the house, she spilled her milk. Spilling milk is what three-year-old children do. I sighed. I did not yell. I simply sighed and went to get a paper towel. Emily was upset and blurted out, "I didn't make the spill!" I asked her who had spilled the milk as I was on the other side of the room. She said, "My hand did it." If I was not trying to get us out the door, I might have laughed at her fairly sound three-year-old logic. Instead, we were delayed as I explained her hand was connected to her body and that her brain controlled her body. So, although I was not angry that she spilled her milk, she needed to take responsibility for the spill and help me wipe it up. When people try to sidestep their responsibility for a situation, I go back to that long ago morning and am reminded that a hand cannot work on its own. A person's brain must tell it what to do.

The second story I share about responsibility and choice involves my son when he was 13 years old. He is a two-time Non-Hodgkin's Lymphoma survivor and underwent chemotherapy treatment between the ages of 10 and 14. Similar to many children diagnosed with cancer, he wanted to be a 'normal' kid. We had many arguments about why he had to endure certain therapies to fight his cancer, and we had arguments about why he had to complete his school work as he had cancer. Looking back, he was pushing his environmental boundaries just as any other adolescent. During this time, he prepared for his Bar Mitzvah ceremony that marks the transition point for him to begin to assume adult ritual responsibilities. Part of his preparation was to write a speech that explained a lesson he learned from the weekly Torah portion. The section he chose to discuss described people who made a choice to become a servant to the priests within the ancient temple. He talked about this section and the responsibilities he would need to consider as he made choices as an adult Jew. He went on to explain, "I did not choose to get cancer, but I can choose how I behave now that I have cancer." Truly, I did not see the speech until after he had written it; however, as the adage goes … out of the mouth of babes. This teenager understood that even though he held no responsibility for the unpleasant situation he found himself, he was fully empowered to take charge of his actions to move forward.

When diverse groups of people are brought together, everyone involved has a choice how they will act. Psychologists talk about fight or flight responses. So it is not unnatural for people to automatically jump into 'us versus them' confrontation mode when they do not yet feel secure within a new work environment even if all parties involved are assigned a common mission. How much energy is wasted putting up antagonistic walls? After much time wasting, we finally let down our guard, loosen our bias, initiate conversations, and then discover we have much more in common than not.

Bruce Tuckman's *Forming, Storming, Norming, Performing* group development model accurately describes most team merging scenarios. My concern is why do we need to spend so much time and energy within the Storming segment? Conversation is an undervalued tool that can help teams move quickly from the Forming to the Performing phase.

Initiating conversation enables people to explore their differences with dignity. It is a means of civil discourse that transforms potential confrontation into understanding and constructive action. Nadia Bolz-Weber, a Lutheran pastor, for The House for All Sinners and Saints congregation in Denver, Colorado discovered the value of conversation to help people from different backgrounds find common ground. She founded her congregation to serve people who were not part of the stereotypical church-going populace. She brought spiritual grace to people who were not comfortable attending mainstream churches. Tattoos and leather jackets were the norm among her young hipster congregants. She found herself in a quandary when mainstream folks wearing collared shirts and khakis began to attend services. She was not sure how to react to these newcomers, because she had launched her ministry to help people who often were shunned by mainstream people. In an odd thought reversal, she did not appreciate these 'others' attending the church she had built to support avant-garde lifestyles. She sought counsel from a friend with a similar ministry who suggested she open her mind. She took the opportunity during a scheduled congregational meeting to discuss the purpose and mission of the church. The conversation that ensued between the two groups focused on what they hoped to gain through their church attendance. The original congregants realized that although they did not necessarily share similar lifestyles with the newcomers, they did share spiritual commonalities and could appreciate each other's differences. The congregation flourished amidst diversity (Tippett 2014b).

CONSIDER AND DELIBERATE

Reflect back on preconceived expectations that turned out not to be as you expected in the following scenarios. Briefly note the object, person, or activity.

New Food—
Outdoor Activity or Outing—
Music/Theater Genre—
First Day at a new Job—
First Meeting with Temporary Assignment Team—

How many of these situations were positive experiences and how many were negative?

Did you need to overcome any extenuating challenges for these scenarios?

What commonalities existed between the varying experiences?

What risk level would you assign for each of these experiences?

In what way does the level of risk affect how you approach uncertainty?

Change is hard for many people. There can be great comfort in maintaining the status quo. If you are trying to initiate a change and find some people are hesitant to embrace your ideas, it is imperative for you to initiate a conversation to better understand their fears and concerns. You may find they did not fully understand your explanation, and once their questions are answered, they become proponents for your proposed action. Or, you might realize their reluctance is justified, and they will join with you to develop a better solution. In either case, being able to facilitate a conversation is a valuable tool that enables all parties to gain an understanding of each other's needs, wants, strengths, and weaknesses, so that they can unify their efforts to achieve their common goals.

Using Conversation as a Gap Closer

Pulling meaning out of conversation is determined by the commonality of language. Language is part of a culture that is defined by word usage, context, and tone. Ben Marcus, an American writer, made an interesting point when he wrote, "There are allegedly two stories to tell: a stranger comes to town, or a person went on a journey" (Marcus 2004, p. xii). When people come together to create something larger than themselves, they are starting a journey. They will need to depend on and cooperate with people who they may not know very well. What is scarier than heading into an unknown abyss with people who you have not yet learned to trust? Conversation provides the space for the language and interaction that will help you stop being strangers and come together to successfully find your way.

When people are brought together to form a working team, they bring a variety of expertise, experience, and jargon. Small misunderstandings can evolve into major standoffs, because people are using words that might have different meaning to different people. Most people want the work effort to be successful; they simply are used to performing the work differently or using different terminology. As the procedure or jargon is outside their comfort zone, people become afraid that they will be negatively judged and shut down. This is why professional organizations such as the Project Management Institute (PMI) and the Society of Human Resource Managers (SHRM) develop a Body of Knowledge that establishes a common language of concepts, methods, and terminology for practicing professionals.

Growing up mixed race in South Africa, Trevor Noah discovered the power a common language can provide when people who do not know each other focus on the obvious differences. He spoke several languages and learned that speaking with other people in their dialect was a tremendous tool to help him avoid trouble. People would speak with their friends not expecting Noah to understand their words, much less be fluent in their own native tongue. Undaunted, he would calmly join in their conversation in his 'simulcast' mode mimicking their exact dialect and intonation. Hearing someone who did not look similar to them speaking with the same inflection and jargon as they did caught them off guard, but within moments they started speaking with Noah as if he was one of their group (Noah 2016).

Conversations and Virtual Teams

The challenge for twenty-first-century leaders is how to align, motivate, and coordinate efforts of people who are not colocated and may never have met face-to-face. Managers and leaders must develop a more complex set of skills to connect and align people who are dispersed across geography and time zones. The differences multiple languages and regional dialects bring to a global workplace are intensified when teams work virtually. No one is right or wrong—just different—until the working group can agree on a common procedure and understand each other's jargon.

Creating virtual connections and strong relationships among dispersed team members is the sign of a strong leader. How do remotely located managers discover, share, and combine the strengths of their reporting teams? How do they motivate their staff and monitor work activities from afar? These work environments require leaders to demonstrate exemplary communication skills. If your team understands the group mission and how their work efforts align with the organization's goals, they will not buckle at the reporting needed to assure milestones and related tasks are synchronized with budget, resource, and schedule expectations.

Several organizations successfully use instant messaging (IM) applications designed specifically to promote informal teambuilding relationships that develop from casual office conversations. A word of caution, just like office interruptions can impede activities that require heavy concentration, too much IM activity can interfere with a person's productivity when they need to fully concentrate on a work task. In my cubicle days, we would post a note asking not to be interrupted. Similarly, an icon or some other notification needs to be identified by your group to realize people are 'heads down' working and not being antisocial.

Virtual team leadership requires a tremendous amount of collaborative activity. A single person cannot create a virtual community by themselves. A virtual community can only happen through the collaborative efforts of everyone involved. Taking the extra time to build trust is a rewarding experience for both personal satisfaction and measurable productivity increases.

CONSIDER AND DELIBERATE

Describe a time when you worked with a group of people who were not colocated.

How did you establish a sense of trust?

How did this group communicate and track work efforts?

What was the most challenging aspect for you as a member of this virtual work team?

Leaders must model effective communications. Patience, persistence, and direct one-to-one connections using a variety of media options, including phone calls, can help to establish effective virtual communication norms. When I worked for an international Enterprise Resource Planning (ERP) Company, I discovered during a conversation with a European colleague how severely behind schedule an announced software release had become. Between us, we worked out a solution in which he borrowed two of my staff resources. The cross-Atlantic collaborative team worked together to reduce the estimated nine month delay to one week. This was a tremendous accomplishment that would never have happened, if we had not worked at developing a virtual relationship. I cannot overemphasize how important a tool conversation is for leaders who need to mediate conflict caused by misunderstandings among team members whether people are colocated or working virtually.

```
┌─────────────────────────────────────┐
│        Message clarity checklist     │
│                                      │
│   ✓  Descriptive header              │
│   ✓  Brief discussion points         │
│   ✓  Required action                 │
│   ✓  Timeframe needed                │
│   ✓  Appropriate attachments         │
└─────────────────────────────────────┘
```

Figure 3.2   Message clarity checklist.

Overcommunicating

Some leaders resort to 'overcommunicating' by sending multiple e-mail messages to dispersed team members. When they do not respond as desired, they get angry and start blasting multiple communications and addressing more people than necessary. If people are nonresponsive to your messages and you have sent a reminder, do not blame your people for bad behavior. Take a moment and review the message you are sending. Is it clear and concise? Do you need to read through paragraphs of verbiage to get to the task assignment or request? Do you use the message header line to succinctly identify the topic? Remember—you are sending a message to people who along with diverse workstyles all have different communication styles. Keep the message simple and brief. Attach necessary documents, share appropriate links, or arrange for a conference call to clarify. Figure 3.2 provides a checklist summary to help assure clear and concise messaging.

Does the e-mail header or agenda title describe the purpose of the communication? What tools did you use to maintain focus and make sure that you will receive the action you desire? Have you provided your staff the background information (e.g., spreadsheets and previous reports) they will need to perform the requested tasks?

Culture, Conversation, and Collaboration

The key to successful communication and conversations is to use nonjudgmental language. An authentic, even tone will help you to clarify facts and intent while maintaining a position of strength and calm. All viewpoints can be discussed and evaluated with the hopes of coming to some type of constructive agreement that enables the group to move forward. Listening skills, respect, and a desire to learn how others view the situation are the keys to creating trust across organization, geographic, and cultural divides.

Culture is an interesting dynamic whether you are collaborating within the same location or across many miles. Growing up, the golden rule that said "do unto others as you would have them do unto you" was drilled into me as the hallmark of basic respect. Recently, I heard someone mention that we should consider treating people how they prefer to be treated. Behavior we consider respectful may or may not be respectful based on someone else's culture, ethnic or religious norms. There may be a brief overly polite *after you, no after you* shuffle as people from different organizational cultures work through the protocols that work best for them; however, the respect they develop through this process will go a long way toward fostering trust and the ensuing collaborative efforts.

The term culture can reference personal workstyle or organizational work environment as well as ethnic or religious norms. Have you had difficulties working with other people but could not figure out why you could not get along? Workstyles have caused conflict since the time of hunters and gatherers. So whether the diversity of your team is based on ethnic, generational, personality, or workstyle traits, leaders need to help their teams recognize each other's strengths and enable them to learn how to come together to create a strong team effort.

Some questions you can ask yourself as you compare yourself to others in your workgroup might help you recognize when you are working within an organizational environment with a compatible culture:

- Are you an introvert or an extrovert?
- Are you an early morning person or do you work more productively later in the day?
- Do you prefer just the facts or do you think like a spider web?
- Do you need to document every conversation or wait to summarize the final decision?
- Do you prefer brainstorming solutions or do you prefer to reflect on a challenge before having a discussion?
- Are you comfortable communicating primarily via technology or do you prefer a more personal touch?

Some people make a generational distinction with regard to workstyles and approach to colleagues within a work environment. However, all of the above personality traits and working styles can be found among baby boomers, millennials, gen-Xers, and so on.

CONSIDER AND DELIBERATE

Reflect on a time when you did not think you would get along with someone, but by the end of the project you had developed a professional friendship.

What were your initial thoughts about this person?

Was there a specific moment that changed the direction of your relationship?

How did you describe that person when the project ended?

Workstyles will change on the basis of your experience, work environments, timelines, and other activities you are juggling in your personal and professional lives. Feedback is an excellent tool to help keep your perceived traits and workstyles aligned with the reality by which your colleagues and team members perceive you.

Conversational Feedback

Conversation is a two-way process that enables people to summarize considered alternatives and explain their thought process. Feedback is a process that begins with someone describing how another person and their actions are perceived. Less than stellar feedback sessions usually end as one-way data streams. When the person providing the feedback allows the receiver to ask questions and explain any misconceptions, the feedback session becomes a constructive conversation.

For many people, the concept of inviting people to share their perspective and provide feedback to your thoughts is a scary situation. Collaborative conversations do not always happen easily. Trust may not have fully developed. People may be concerned about sharing information among teammates with whom they have never worked before. Or, they may be worried about sounding 'stupid' to new colleagues. To encourage productive collaborative sessions among newly formed groups, a facilitator can help the group explore possibilities. The Collaborative SCHOLAR Conversational model illustrated in Figure 3.3 is a facilitation tool that can help the working groups explore boundaries and options which may not otherwise be considered.

The SCHOLAR acronym identifies a seven-step process to facilitate innovate problem solving. Each of the seven steps takes your team from an initial conversation that defines a common understanding of the starting point, the solution development phase, and the change management procedures. Reality may cause you to

Scrutinize	Strengths/weaknesses of current reality
Connect	Reach out through the organization
Hone	Short- and long-term vision
Options	Brainstorm, review, and evaluate
Leverage	Available assets
Accountable	Own your actions
Refine	Adaptation and improvement

Figure 3.3   Collaborative SCHOLAR conversation model.

need to refine the solution plan. Simply, use the collaborative process to retool and adapt the original plan to create the final solution.

The Scrutinize phase encourages your working group to analyze the current situation. Document the assets you have working in your favor and document those challenges that could impede your progress. The Connect phase encourages reaching out into other areas of your organization that might have needed expertise or experience. The Hone phase suggests you break your vision into short-term and long-term goals. Being able to introduce smaller successes may help other people in your organization to gain confidence and more fully support your ultimate purpose. The Options phase is a time to think about alternative ways to accomplish your goal. Once people start to review and evaluate alternatives, some ideas may be discarded; however, an idea that has never before been deployed may be the launch point for an award-winning solution. The Leverage phase encourages people to review existing resources that are underutilized or can be repurposed.

The final two steps are the most important for leaders to remember. Be accountable and own your actions. Share all of the positive credit with your group members. When the work effort provides a less than positive experience, retain the misstep within your realm of responsibility. Share the lessons learned as feedback so that you and your team will not make similar mistakes in the future. Lastly, be aware that nothing is permanent. Environmental conditions will change, and you will need to refine your approach. Be ready to help your team learn from mistakes, adapt, and continue forward.

Summary

The Power of Collaboration and Conversation

- Strong leaders help collaborative teams identify their unique capabilities and integrate the efforts of all the participants into a collaborative force to achieve the shared vision.
- A measure of successful collaboration is defined by how teams are able to develop collective responsibilities to adapt inherited cultures to meet the needs of the evolving community.
- Conversation is a tool that enables people to avoid confrontation and began to gain an understanding of each other's needs, wants, strengths, and weaknesses to unify their efforts to achieve common goals.
- Collaborative conversation is a powerful process that can be used by diverse working groups to promote innovative problem solving.

Chapter 4

Building on Strengths

Matrix teams enable ...
Different workstyles are ...
Collaborative workspaces assure ...

Pierre Teilhard de Chardin, a French Jesuit and paleontologist, prophesized, "The coming age of evolution won't be driven by physical adaption, but by human consciousness, creativity, and spirit." The astonishing aspect of this quote is that he was born at the start of the scientifically driven industrial revolution and lived through the development of major twentieth century inventions: motor cars, radio, airplanes, and so on. A common value of this era was that machines would drive the progress of man. Yet, he puts his confidence in the capabilities of man.

De Chardin began his scientific career in China, where he worked with the team that discovered the Peking Man. Through his studies, he recognized the role physical adaptation and evolution played in man's early development. Midcareer, he switched his studies to philosophy to explore modern man's thought processes. He was especially interested in how the sharing of ideas could help evolutionary adaptation. He did not believe innovation could occur within a vacuum. He proved through his research that group discussion was the core foundation responsible for human survival. He visualized a cyclical thought process which centered on people deliberating and debating ideas. One person builds on another person's ideas until a practical reality was created that might never have been practiced before (Tippet 2014a).

In this same vein, leaders identify and leverage the individual strengths of people with diverse workstyles to create something bigger than they could have created on their own. Team leaders, who are able to recognize the varying experiences that different people bring to a challenge, are already several steps ahead of the manager who assumes there is only one way to resolve an issue. Success is dependent on a team's ability to foster collaborative group discussions that enable all members to share, comment, and adapt ideas. A single loud voice insisting there is only one way to resolve the situation only considers the perspective of one person. Collaborative discussions incorporate a variety of viewpoints which manipulate, twist, and reshape thoughts to create an innovative solution.

CONSIDER AND DELIBERATE

Describe a situation in either your personal or professional life when you and other people were stymied by an unexpected challenge.

What caused the process/project/activity to stall?

How did the situation become *unstuck*? Did a single person save the day or was it a group effort?

Could the problem situation have been prevented? How?

What lessons were learned from the situation?

The best ideas always seem to surface when the group has run out of standard solutions. People begin to toss around impossible suggestions and suddenly some-one will say, "Wait that just might work." The group then begins to refine the out-landish suggestion until a workable solution is finalized.

Collaborative Working Groups

Working relationships among team members will adjust over time taken, as people try out different voices and take on a variety of leading and supporting roles to pur-sue various collaborative efforts. Assignments will broaden as people discover skills and expertise they might never have realized that they, or others, possessed. These unofficial roles may be assumed by people who are particularly good at network-ing throughout the organization, those who plan celebrations, those who simplify complicated situations, or those who are particularly good at note taking, editing, or fact-checking. Leaders need to pay attention to these shifts by adjusting motiva-tors, recognition, and success metrics to mirror the expanding productivity of the group and individual team members.

The Corporate Executive Board (CEB), a best practice insight and technology company, initiated a global, cross-industry study of 1440 front-line service repre-sentatives in 2015. The data culled from the study identified seven discrete person-ality types to describe the people who performed contact center functions, as well as the associated level of effectiveness for each personality type who performed as a help-desk service rep. The seven personality types identified by the study are as follows: (1) The Controller, (2) the Rock, (3) the Accommodator, (4) the Empathizer, (5) the Hard Worker, (6) the Innovator, and (7) the Competitor. As might be expected, the largest group of individuals (32%) identified as an Empathizer that is described as someone who is an empathetic listener. However, the people who identified as a Controller (15%) were recognized as providing the most efficient and pain-free customer service. The report describes a Controller as someone with strong opinions and who likes to demonstrate their knowledge. Basically, they are skilled at confidently telling people what to do and how to do it—and they enjoy doing so (Dixon et al. 2017, pp. 111–115).

The results from the CEB front-line service representatives study are not so surprising with regard to the current focus of customer-service organizations. The data's value is that it demonstrates a microcosm of the anecdotal collaborative team referenced throughout this book. Table 4.1 defines each of the seven roles as defined by the study. The rightmost column is included so you can reflect on the various representations of each personality type within your organization team. Additional rows are inserted for you to use to identify any additional personality types that may be unique to your team.

Every working group is made up of a blend of the roles identified in Table 4.1. Depending on the organizational function performed, the percentages of the people who represent the different roles will vary. For example, solution-based sales teams will have a significant percentage of people who identify as a Competitor. However, this does not mean that only Competitors work in a sales-related function. Every sales team is made up of people who identify

Table 4.1 Working Group Roles

Call Center Representative Team Role	General Behaviors	Corporate Executive Board Study Percentage (%)	Your Organization
The controller	Likes to share opinions and direct group activities	15	
The rock	Maintains optimism and calm in all situations	12	
The accommodator	Brings many voices into decision and works for people to meet halfway	11	
The empathizer	Listens carefully and enjoys helping others resolve issues	32	
The hard worker	Insists on following procedures and adheres to deadlines	20	

(Continued)

Table 4.1 (*Continued*) Working Group Roles

Call Center Representative Team Role	*General Behaviors*	*Corporate Executive Board Study Percentage (%)*	*Your Organization*
The innovator	Enjoys creating new methods to enhance established procedures	9	
The competitor	Focuses on goal achievement and views activities as win/lose opportunities	1	

and perform the function of one of the described roles. Collaborative teams are most successful when the various members are able to analyze a situation from different perspectives.

CONSIDER AND DELIBERATE

Looking back on the past three years, list the different professional and personal groups for which you have been a member. Next to each group, identify the role(s) listed in Table 4.1 that you played within that group environment.

To the best of your recollection, how were the seven roles represented among your colleagues over the course of these different group experiences? Was there a significant percentage of one type of role or an even distribution?

Which group experience did you enjoy the most? Why?

Which group experience did you enjoy the least? Why?

Did you perform the same functional role(s) during each?

Out of all the roles you listed, which do you find naturally you gravitate toward?

People will take on different roles when working within different team environments and may even shift their participatory role on longer term projects as they gain skills and take on responsibilities of people who move on to other projects. No role is *better* than another role. Each one contributes to the total effort. In fact, it might prove beneficial to encourage team members to switch their natural roles within the team for a day to encourage broader perspective and innovation.

Culture and Collaborative Groups

Colonel Scott A. Jackson shared an interesting U.S. Army cultural tradition that follows the absorption process of the entry-level lieutenant into a fully operational working unit. This entry officer level often oversees noncommissioned officers who have many years of experience. A young officer must prove he is worthy of the soldier's respect. The Lieutenant title does not guarantee a young officer automatic respect. The orientation process begins with the announcement that 'a' lieutenant has been assigned to the unit. When the young officer arrives he or she is referred to as 'the' Lieutenant. Senior officers know that the young officer has established trust and become fully integrated into the unit operations when the reporting soldiers refer to her or him as 'our' Lieutenant (Colonel Scott A. Jackson [personal interview, May 26, 2017]). I had a similar experience working with my seasoned technicians at AT&T. The day when they stopped calling me "Susan" and started calling me "Schwartzy," I knew I was considered part of their team as they only called their trusted team members by their last names. Giving me a nickname was their way of telling me that I was doing more than okay.

The culture of trust within the group determines the roles that different members play. Often the culture described by organizational leaders may not actually be the culture people experience day to day. Edgar Schein, a noted scholar of organization development, observed that in every organization, there are three different cultural levels (Figure 4.1). The artifact's layer is made up of promotional images

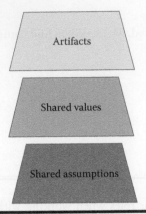

Figure 4.1 Cultural organizational levels.

that define the organization's publicly seen cultural posture. These can be legends passed down from one employee generation to another or a visual image such as the building's architectural façade. Other artifacts might include statues, paintings, or visualizations of specific phrases or quotes representing the founding principles that contribute to the culture.

The second and third levels of Schein's model are about where the actual working culture of the organization is discovered. The second Shared Values level represents what the organization says they are or would like to be. This is the level that contains the strategies, goals, and philosophies that provide the foundation for the organization's business practices.

The third Shared Assumptions level is what the people within the organization experience on a daily basis. Schein describes this level as being where the unconscious perceptions of the organization's management and staff are found. This is the actual working culture of the organization (Schein 2004, pp. 25–37).

An example of the actual working culture of an organization can be an unexpected positive or negative. Have you ever discovered that someone described as a difficult client or supervising manager had qualities and knowledge that others did not appreciate? The dreaded assignment turned into a golden opportunity for you to learn and build skills. On the other hand, have you ever accepted a position only to discover within a few days of your hire that your actual duties are different than what was described in the position description and interview process? Were you ever told you were being hired for your skills and creativity only to discover that the only ideas put forward belong to your supervising manager who issues declarative orders and does not entertain any collaborative discussion regarding his/her decision? When you query your team members for counsel because you believe you discovered a problem, do they tell you to keep quiet and do as you're told? In this final instance, it does not matter what the company logo or strategic plan espouses; the culture of a working group is determined by the Shared Assumptions the team members hold.

Creating a Collaborative Environment

When people are afraid to speak up because they are afraid of being belittled, or possibly punished, they will nod in agreement even though they may be aware of essential information that should be considered. They simply want to survive unscathed from any retribution from the person who wants to control the discussion and group outcome. Innovative ideas that could decrease costs, improve operations, or increase sales are lost because people are too fearful of the repercussions that occur when people disagree with the strongest voice in the room.

James Detert and Ethan Burris identified common management pitfalls demonstrated by even the best intentioned leaders that can prevent people from speaking up when they have concerns. These challenges are categorized within two areas: fear factors and futility factors. One practice they identified as fear factor inducing is the practice of relying on anonymous feedback. Although industry best practices have used anonymity as a means to encourage open and honest feedback, it actually inhibits frank discussions that can lead to problem resolutions. Instead of encouraging people to bring concerns forward, these anonymous suggestion boxes are actually sending the subliminal message, "It's not safe to share your views openly in this organization."

The second cause for concern regarding the practice of anonymity they discovered is that when negative feedback was channeled via the suggestion box, it was not uncommon for management to pursue a debilitating witch hunt to determine who perpetuated the negative feedback. The employees in these organizations will definitely think twice before they place any future negative feedback into the comment box.

Their third concern with using anonymity practices is that, without knowing the source of concern, people may take the original premise out of context that can create a larger problem as the message originator is not available to more fully explain their comments. Or, in the case of anonymous messages reporting abusive behavior, appropriate measures cannot be taken to rectify the situation as the HR Ombudsman cannot initiate action without being able to speak with someone to start the formal exploratory process. If the accusation is specific enough, the manager will recognize the person and possibly take further abusive action.

Other instances that might be categorized as a fear factor inducer are the result of managers who say they have an open door policy but will not consider any comments or feedback regarding projects for which they are the lead or heavily involved. Body language may also prevent people from sharing feedback. The authors suggest managers be cognizant of 'power' postures. For example, leaning back from behind a large desk may seem to present relaxing environment to staff; however, it establishes a power divide—more similar to being on opposite sides of a stream that does not encourage collaborative, peer-based conversation. Gathering around a table where everyone is sitting in similar chairs will promote more open discussions.

When people take the time to provide input, but then it is not included in the final report or presentation without explanation, they are being told that their efforts are not meaningful. There could be very good reasons the information was not included, but if these folks are not provided an explanation, they may not be quite as willing to provide support in the future. This is what the authors mean by a futility factor. Sometimes people will say that what they received was not what they asked the person to prepare, there was no time for correction, and so they could not include the person's work. Managers need to assure that they provide people specific directions and, if possible, examples of the desired format, to eliminate any uncertainty regarding deliverables. Lastly, the authors caution managers about asking people to explore new ideas, if there is no possibility that funds will be made available to implement the findings. These futile efforts will be seen as busy work that distracted them from more productive activities.

Managers and leaders can encourage a constructive feedback culture by incorporating feedback as part of daily casual conversation. People will become accustomed to sharing ideas without feeling as if they are being put on the spot. Other suggestions that the authors mention include being transparent when priorities or budgets change. You may not believe it is appropriate to share the specific details that went into the necessary change, but a certain amount of transparency will help one to mitigate any futility that people might be feeling. By sharing the challenges that came up during executive discussions, they will believe that you did your best to present their viewpoints and concerns. Whether the action is about a decision not to include a chart in a presentation or to stop a significant work effort, managers and leaders who take the time to close the loop with a constructive conversation are letting the people on their team know that they and their work are appreciated and valued (Detert and Burris 2016).

CONSIDER AND DELIBERATE

Select an organization you have worked within professionally or as a volunteer. Briefly outline the culture based on the Schein's three-level model: (1) artifacts, (2) shared values, and (3) the shared assumptions.

Do the three levels align? If not, where did the disconnect occur?

How did the alignment or nonalignment affect your experience?

Based on your reflections, how did the leaders of the group inhibit or promote fear and futility factors among the working group members?

The most successful organizational leaders work hard to assure that the publicly promoted cultural branding aligns with the execution of the internally professed business practices, values, and operational procedures. When these three tiers align, people are encouraged to assure that the smallest operational details align with the organization message and mission resulting in a positively energized workplace.

The Culture of Matrix Teams

Matrix organization structures are often used as the first step to integrate previously separate organizations. This first step toward a potentially larger organizational change offers a leader an enormous opportunity. The challenge is how to align the current cultures of the working groups that are being brought together and the desired culture which is also known as the "where we want to be." A common management error is to ignore the differences and assume that everyone will immediately adapt and wear the new corporate uniform with pride. People are not robots. It is part of human nature to want to have some control over our destiny. The most successful change and merger efforts are driven from within the affected working groups by people who identify as part of the organization. Leaders did not necessarily drive the change; instead, they empowered individuals to create the change needed to get their groups to the desired destination.

The first steps to organizational integration often begin with cross-functional teams that are created by the overseeing managers *borrowing* people from other units within the organization. The benefits of *borrowing* people include the opportunity to introduce a variety of experiences and perspectives that can help one to drive innovation while minimizing costs and redundant operations. However, the dynamics of merging different work groups can be difficult. In the case of conflicting priorities, with whom will staff members align themselves—the *dotted-line* manager or the person who completes their performance review? In brief, matrix team leaders have all of the responsibility without any of the authority.

Balancing the complexities of a matrix team environment is a practical challenge for everyone involved. The team leader must corral diverse personalities, multiple reporting structures, and varying performance expectations to develop a synchronized group of people. Successful matrix team leaders need to be able to keep everyone's focus on the alignment of the group and organizational goals. They must recognize different workstyles, minimize competing objectives, and synchronize resource coordination and communications. If they can achieve this delicate balance, the broader perspectives and diverse experiences fostered by matrix team alignments can provide great value to the organization and participating individuals.

Diversity of Communication Workstyles

Many workplace difficulties are rooted in simple misunderstandings. When people take the time to work through interpersonal conflicts, the ultimate cause is most often attributed to different communication styles and expectations. Being able to recognize and adapt to other people's communications needs is an essential skill for all people, especially those who work within a matrix organization environment.

There are a number of workplace assessment and profile tools to help teams understand each other's workstyles with the intention of improving communications and productivity. Some of the most popular of these assessment tools are Myers–Briggs Type Indicator®, EQ-i 2.0®, and DiSC®. The DiSC profile categorizes people as one of the four types: (1) Driver, (2) Influencer, (3) Steadiness, and (4) Conscientious. Table 4.2 addresses the traditional DiSC identity profiles from

Table 4.2 DISC Communication Style Profiles

	Driver	Influencer	Steadiness	Conscientious
Values	Competency Concrete results Action	Coaching Counseling Freedom of expression	Loyalty Helping others Security	Quality Accuracy
Motivation	Winning Competition Success	Social recognition Group attributes Relationships	Cooperation Opportunities to help	Opportunity to gain knowledge Show expertise Perform quality work
Priority	Accepting challenges Taking action Achieve immediate results	Taking action Collaboration Express enthusiasm	Giving support Collaboration Maintain stability	Ensure accuracy Maintain stability
Attribute	Direct, demanding, forceful, determined, fast	Convincing, magnetic, enthusiastic, trusting, optimistic	Calm, patient, predictable, deliberate, stable	Careful, cautious, systematic, diplomatic, accurate
Limits	Impatient Skeptical	Impulsive Disorganization	Indecisive Overly accommodates Tendency to avoid	Overly critical Over analyze Isolating

(Continued)

Table 4.2 (*Continued*) DISC Communication Style Profiles

	Driver	Influencer	Steadiness	Conscientious
Fears	Being seen as vulnerable	Loss of influence Disapproval Being ignored	Loss of stability Change Offending others	Criticism Being wrong
Communication requirements	Bottom line Be brief and focused	Share experiences Ask questions	Be amiable Define expectations Time for clarification	Focus on facts Minimize "pep" talk Be patient and persistent
Needs others to:	Weigh pros and cons Calculate risk Research facts Deliberate Recognize others' needs	Focus on facts Speak directly Have systematic approach Demonstrate follow thru	React quickly to change Multitask Apply pressure to others Initiate flexibility Thrive on unpredictability	Delegate tasks Make quick decisions Compromise Encourage team work

a communications perspective. If you are not familiar with DiSC profiles, the first four rows will help you identify the different workstyle tendency for you and your colleagues. The lower four rows compare the communication preferences and needs for each profile.

By understanding your coworkers' limitations and fears, you can provide the supportive tasks identified in the final row. You may even find that your impossible boss may not be quite so impossible, if you are able to adjust your communication style to best meet their needs.

Personally, I am an Influencer and think like a spiderweb. I work through all kinds of scenarios and want to share them all. Over the years, I have worked for several managers who would be identified as Drivers. I was in my 20s when I worked for my first Driver manager. It was not a positive experience for me. Reflecting back, I expected him to adapt his style to support me. It was an immature viewpoint and I paid a political price. Midcareer, when I encountered another Driver manager, I had become a manager and a parent and was able to view situations from a broader perspective. I recognized my manager's need to be presented the essential decision points while being prepared with backup data to answer questions. My work life became so much easier once I understood how my manager's workstyle differed from mine and I was able to adapt my workstyle to best support my manager.

CONSIDER AND DELIBERATE

Referencing the DiSC communications style table above, which workstyle do you believe best describes you?

Think about someone who you would characterize as a difficult person. Which workstyle do you think best describes them?

How can you adapt your behavior to change the dynamics and hopefully improve your working relationship?

As already mentioned, matrix teams require leaders to operate via a multidimensional lens. People may report to several managers who will have different communication needs. The varying workstyles of your peer colleagues and reporting staff insert further complexity. Whether your organization is a small business, global corporation, government entity, not-for-profit, or a volunteer community group, people do make the difference. Paying attention and responding appropriately to varying workstyle and communication needs is an effective means to motivate people to perform at their highest capabilities.

Workstyles and Team Tasks

People may have individual workstyles; however, individuals will perform a variety of tasks within a team environment. Sometimes one person may take the lead and at other times they will focus on completing a specific task. Leaders need to watch closely to make sure that each team member takes on task types that will leverage their capabilities, incorporate professional growth opportunities, and move the team toward the designated goal.

Meredith Belbin, an industrial psychologist, developed a Team Role theory that organizes team functions into three task categories:

- Action-oriented
- People-oriented
- Thought-oriented

Action-oriented task roles represent the doers who work to complete the specific work activities to achieve the goal. The people-oriented task roles represent the person within the group who is well-networked and always seem to know someone who knows someone who could provide information or some other kind of assistance. Thought-oriented task roles represent people who are able to provide the group a view of the larger picture such as subject-matter specialists, quality monitors, and resource managers. Most people will perform all three types of task roles at different times depending on the project and other team member skills.

Expanding Belbin's theory to a leadership perspective, a person might find him/herself taking on a leadership type of task role even when they have not officially been designated the leader. Or, designated leaders may find themselves performing a specific worker-bee type function. Flexibility is a strong attribute for leaders to possess. Each task that builds toward achieving the designated goal is important. Egos cannot get in the way. If a task needs doing and everyone on the team is constructively engaged, roll up your sleeves and attend to the requirement at hand. When your team sees you pitching in to help with mundane jobs, they quickly see the importance of what they might have considered a menial task and are more motivated to work harder to achieve the shared team goals.

An example of just such a situation involved my son, Carl, who was training to be a chef with Executive Chef Bryan Voltaggio at his fine dining restaurant Volt in Frederick, Maryland. This was the summer when Voltaggio opened his first Family Meal restaurant. Although Carl had the late shift which involved cleaning and closing the Volt fine dining kitchen, he had been told to be at the Family Meal kitchen early to deep clean the refrigerator in preparations for the new restaurant opening. Needless to say, Carl was not a happy worker. When he reported to work in the morning, the only other person in the kitchen was Bryan Voltaggio, the executive chef and owner. The two of them worked together to get the job done. Carl realized that deep cleaning the refrigerator was essential to running a restaurant, and not a menial task to assign to the most junior team member. The conversation they shared while they scrubbed offered my son a unique mentoring opportunity. To this day, Carl compares his head chefs and managers by the yardstick he established watching Voltaggio demonstrate leadership by example.

CONSIDER AND DELIBERATE

Consider a recent work effort you performed as part of a group. What tasks/responsibilities did you perform as part of the work effort?

Can you assign each of these tasks/responsibilities to the appropriate category?

Action-oriented:
People-oriented:
Thought-oriented:

Select another work effort you performed as part of a group. What tasks/responsibilities did you perform as part of the work effort?

Can you assign each of these tasks/responsibilities to the appropriate category?

Action-oriented:
People-oriented:
Thought-oriented:

How did the roles you perform during the first work effort compare with the roles you performed as part of the second work effort?

Which roles did you enjoy performing most? Why?

Shifting between action, people, and thought-oriented task roles provide teams many benefits. Exercising different working muscles and broadening perspectives will help team members develop new skills and establish stronger relationships with other team members which will enhance productivity for the entire work group.

Coaching for Performance

Some organizations measure a leader's effectiveness by the quality of the people they develop and succession planning that has been put into place. When a senior leader brags about their indispensability to the organization, it really means they have not trained anyone to take on his/her responsibilities. Yes, they are indispensable; so indispensable in fact, they cannot be considered for promotion as they had not trained a successor.

Dr. Henry Cloud defines a leader's role as "the creation of the kinds of conditions in which people can bring their brains, gifts, hearts, talents, and energy to the realization of a vision" (Cloud 2013, pp. 42–43). Strong leaders are lifelong learners and provide opportunities for people working with them to do so. They are the people who help others realize that they are part of the whole that is greater than the immediate cubicle walls.

When I think back to the leaders who most influenced me, I recall the moments when they helped me to think through the challenge that needed resolving. At that moment in time, I just wanted them to tell me what to do. However, coaching me to identify the *real* issue and work through the possibilities for resolution was a great gift. It might have been easier to tell me what to do; but, taking the time to coach me through the decision process was the right thing to do.

Jorgen Vig Knudstorp was asked about the strongest leadership lesson he learned during his 15-year tenure as Lego's Chief Executive Officer. He responded, "Thank you for doing all the things I never told you to do. You shouldn't run a company based on what you tell employees to do. You should run it based on intention."

He clarified his comment by explaining "context-setting was more important than controlling." His strategy was to outline a vision and create an environment that empowered his staff to take the needed action to achieve the vision. The success of this strategy is demonstrated by the company's turnaround effort that took Lego from a negative cash flow position to the leading global toy manufacturer (McGregor 2016).

The following checklist may prove helpful as you work toward creating the conditions that will empower your team to pursue the needed action to achieve success:

- Sustain mutual respect and purpose
- Be open to hear the truth
- Assure safety of conversation
- Listen for multiple perspectives
- Ask exploratory questions to clarify
- Enhance relationships

Each of these items will elicit a battery of questions that you must answer as it pertains to your workplace environment. The following are a few examples that may help your thought process:

- *Sustain mutual respect and purpose*
  Do you and your team members understand how your assignment aligns with the overall organization mission?
  Are you modeling respectful behavior? How are you handling disrespectful behaviors on your team?
- *Be open to hear the truth*
  Do people only share positive feedback with you?
  How is negative feedback or other examples of staff concerns handled?
- *Assure safety of conversation*
  How often does your team engage in casual conversation?
  Are you open to constructive feedback?
- *Listen for multiple perspectives*
  Do you discuss potential consequences of an action with the people who will be affected by your decision?
  When was the last time you considered a challenging situation from more than your personal perspective?
- *Ask explanatory questions to clarify*
  When people come to you for advice, do you immediately provide an answer or do you ask questions to confirm your understanding of the request?
  During your last team meeting, how many times did you respond to a query with a clarifying question?

■ *Enhance relationships*

When was the last time you had an informal coffee or lunch with members of your team?

When was the last time you had an informal coffee or lunch with a colleague from another department or organization?

Whether you have a formal leadership title or are someone aspiring to be the best professional you can be within your field, the suggested questions can help one to build rapport and trust. Once you have established this base within your working group, people will mirror your positive behaviors. The result of these efforts will be a positive collaborative environment that merges individual strengths to leverage the unique perspective that each team member brings.

Summary

Building Strengths

■ Innovation cannot happen in a vacuum. Together, people build on each other's ideas to envision a reality that might never have been practiced before. Human empowerment drives technological advances, not vice versa.

■ The most effective working groups are made up of people who represent a variety of roles to provide different perspectives and support functions.

■ The culture of trust within a group is shaped by the public organizational image, the espoused values and business practices, and the unconscious organizational assumptions.

■ Constructive conversations and regularly scheduled feedback sessions help people realize their work is appreciated and valued.

■ Successful matrix team leaders need to be able to recognize different communication and workstyles, minimize competing objectives, and synchronize resource coordination and communications to maintain alignment of group and organizational goals.

■ Leaders need to make sure that each team member takes on different task roles to leverage their capabilities, incorporate professional growth opportunities, and move the team toward the designated goal.

■ Leaders need to be able to create conditions that will empower their team to pursue the needed action to achieve success.

Chapter 5

Focus and Motivation

Matrix teams enable ...
Different workstyles are ...
Collaborative workspaces assure ...

Patience and Perseverance

Collaborative conversation and constructive team-building among diverse people take time and can meander along a variety of paths before a resolution is achieved. A leader's role is to pay attention to the process to assure people stay motivated and on track.

Working collaboratively with people from diverse backgrounds can be a great deal of fun. The challenge for leaders is being able to maintain your vision, meld their strengths, and focus their energies. This is the part where working together to achieve a greater whole can sound similar to an unattainable endeavor. It requires a great deal of attention and energy. Leaders must constantly monitor all available feedback channels, be attuned to trending changes, mediate conflict, and enable adaptations to empower teams toward success. More often than not, leaders will need to draw on reserves of patience and diplomacy—multiple times each day—as they assure people stay motivated to achieve what needs to be done.

Leadership—An Ancient Tale

Helping people stay motivated by (1) focusing on what is important, (2) inhibiting activities that are not valuable, and (3) keeping everyone on track is a centuries' old challenge for leaders. There is a wonderful story that provides a simple example of motivational leadership during a very confusing moment in history. During ancient times, the Babylonians conquered Jerusalem, destroyed the Holy Temple, and the conquering generals ordered the Israelites to prepare for the journey to Babylonia where they were to be taken as prisoners. The Temple High Priest encountered several men pulling large stones from the rubble and queried their purpose. The men proudly reported they were planning to take the stones to Babylonia to rebuild the temple so they would have a place to pray while in exile. Instead of congratulating their efforts, the High Priest chastised them. He told the men that the stones were much too heavy to drag on the arduous journey ahead; instead, they should be gathering food and clothing to support their families. The men asked, "How will we pray without a temple?" The High Priest told them as they would be traveling for many weeks and did not know where, or how long, they would be resting along the way, they would never know if there would be an enclosed structure to use for a house of prayer. In lieu of brick and mortar, he told them a sacred space would be created when a minimum of 10 men gathered to pray. Similar to our modern day virtual teams, the Israelites were able to establish community norms and behaviors wherever their wanderings took them. Just as in ancient times, conference rooms and cubicles are not a requirement for people to be able to work together. An understanding that enables the convergence of cooperative human spirit is all that is needed to create a fully functioning community.

Even without the aid of modern day leadership guidebooks, the High Priest (1) focused the men's attention on the priority task of gathering survival essentials, (2) stopped the men wasting energy trying to drag heavy stone pillars across the desert, and (3) helped to keep them on track by providing an innovative solution to their concern that they would not be able to pray.

Human and Organization Needs Satisfaction

The temple exile story demonstrates that the leadership trait of helping people to focus on the basics for survival before attacking a more complex goal is an enduring trait. A more modern research model is Maslow's hierarchy of needs. During the mid-1940s, Dr. Abraham Maslow identified a five-level model, which identifies the human needs that need to be satisfied before *higher level* activities can be performed. Maslow's hierarchy of human needs often is represented by a triangle or pyramid shape. The hierarchy of organization needs and communications model, illustrated by Figure 5.1, merges Maslow's five levels of individual well-being that must be achieved before a person can move to a higher level with the hierarchy of needs that an individual needs to address as they become part of an organization. The organizational communications requirements are melded into this model using the journalistic lens of Who, What, When, Where, and Why to frame the messaging distributed among the various employees, associated stakeholders, and customers.

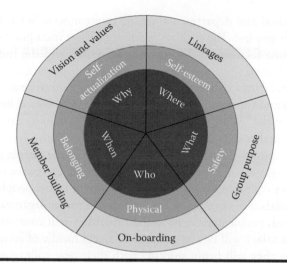

Figure 5.1   Hierarchy of organizational needs and communication.

Maslow's first layer begins with basic human physiologic needs: oxygen, food, water, sleep, and so on. Once these are assured, individuals move their focus up to the safety, belonging, and self-esteem layers. Only after basic survival, essential connections, and core self-esteem elements have been achieved by an individual, can that person begin to work toward his/her optimal potential.

The parallel communications needs model begins with a focus on the individual (the Who) and the information a person needs to become an effective member of the organization. Some people might refer to this segment as on-boarding or the basic survival information a person needs to become an employee. What are their assigned responsibilities? Where do they sit? How and when do they get paid? Where are the restrooms? The focus is on an individual's needs. Once a person feels confident, their individual needs are addressed; the next step is to integrate them into their immediate workgroup (the What).

The *What* segment of organizational communications targets information that describes the purpose and associated procedures of the work team. It consists of tactical information regarding how specific processes and work assignments will be affected by a change. The third Member Building segment (the When) addresses professional development for individual growth and enhanced team-performance strategies that enable the group to execute the scope of their responsibilities within schedule and quality expectations.

The fourth and fifth levels address organization-wide communications. The fourth segment (the Where) addresses how the various groups or departments will share information between their work efforts. The purpose of this segment is developing a message that helps people across the organization to fully understand how their individual and departmental efforts interconnect. What information is needed by which groups? The fifth segment (the Why) incorporates the strategy that keeps everyone focused on the organization vision and the important values that support the vision.

There are similarities between the hierarchy of organizational needs and the communications and the organizational clarity model (Chapter 2) because the strategic messaging or the "Why" is what creates the glue that links the organizational groups/entities together to jointly achieve the overall mission.

Leaders need to understand the primary concerns of the people for whom they are crafting messages to assure the people receiving the message are ready to receive it. When it comes to leading teams through change or crisis, people should focus first on the essentials. For example, when a merger, or major organizational restructure, is announced, people need to be assured that they will continue to have a job and their current salary will not be reduced. The next hurdle of concern for people is when and how they will begin working with their new colleagues to develop a team environment. Once people are secure and confident of their place within the organization, they can then begin to establish linkages over which they channel their expertise and creativity to work with new and existing colleagues to help promote the new organizational mission.

CONSIDER AND DELIBERATE

Describe a situation when you felt as if the proverbial rug was pulled out from under you.

What was your initial reaction?

What was your second thought?

What was the process you used to work through this challenging situation?

What lessons were you able to take with you from this experience?

Reflecting back to a situation when you were not entirely comfortable will help you to prioritize your needs and focus your motivational energies to help others through challenging transitions. You will be able to help your team and associated stakeholders perform to their best ability, because they know the survival essentials are handled and trust the group of people with whom they are working.

Motivational Satisfiers and Dissatisfiers

Much research has been funded trying to understand what makes people do what they do. Some people self-motivate, whereas others need an external push. Managers may use a variety of positive or negative motivational tools to direct their teams toward a desired end. The most effective methods will depend on the motivator, the environment, and the person they are trying to encourage. There is no simple answer when it comes to learning what motivates different individuals. Leading teams is similar to the experience of a parent when they discover that second and third children react differently to similar experiences than their first born. Even though they shared the same parents and grew up in the same household, each child in a family has their own physical and psychological makeup. It is natural that they will have different needs and desires. When you bring people together in a workplace with different cultural and professional experiences, of course, not every motivational-management technique will be successful for all situations. This section introduces several theories that may help you understand and navigate different situations that you might encounter.

During the 1950s, Frederick Herzberg developed the Motivation Hygiene Theory that divides motivational practices for workers into two categories: satisfiers and dissatisfiers.

Tangible items such as salary increases and workgroup assignments that might be used to entice workers to continue with the organization are referenced as dissatisfiers. These 'carrots' when used to encourage dissatisfied workers seldom result in keeping the workers happy. The pay raise may keep the worker motivated for the short term, but at some point, he or she will become unhappy again and leave.

Satisfiers are defined as actions associated with responsibility, growth, and recognition. Satisfiers are less tangible, but assure happier, longer tenured workers. Satisfied workers appreciate salary increases, but they stay on because they feel that they are important to the achievement of the organizational mission.

Moving the Elephant—A Motivational Metaphor

In the book, *Switch: How to Change Things When Change is Hard*, Dan and Chip Heath describe humans as being motivated by emotion and rational thought. They use the visual metaphor of a Rider on top of an Elephant to demonstrate how people are motivated by a variety stimulus that balances the emotional and rational sides of the brain. The Rider is associated with the rational side of the brain and is most effectively motivated via clearly communicated explanations. The Elephant is associated with reactions described as emotional or "gut" reactions. The Elephant and the Rider provide counterpoint such that if a Rider does not have a clear vision of the route, he could lead the Elephant in circles. Moreover, if the Elephant does not want to move in a certain direction, no amount of logical urging from the Rider will get the Elephant to move forward. There is a third element to this model the Heaths refer to as Shaping the Path. This concept provides helpful guides that make the journey easier by encouraging the Elephant and providing clarity to the Rider.

The Elephant–Rider-Path metaphor can be deployed even if you do not have supervisory power or a huge budget. This flexible methodology enables you to address different working personalities and helps your group toward success. The simple checklist reminds you to

1. Motivate the Elephant with a compelling reason.
2. Direct the Rider and establish clarity of purpose.
3. Shape the Path to assist the current situation.

This three-phase thought process allows you to adapt a group's motivation as the situation and requirements change. The one consistent consideration needs to be the end game that everyone is striving to reach together (Heath and Heath 2010).

CONSIDER AND DELIBERATE

Do you think your motivational needs tend more toward being a Rider or being an Elephant? Briefly describe a situation in which you had not *bought into* the purpose until the leader ascribed to your rational or emotional motivational needs.

Does your motivational requirement (Rider vs. Elephant) change depending on the situation? Can you briefly explain the circumstances?

Was there a time when you or the team leader needed to shape the path to help your team maintain their motivation?

Reflect on a time when changing circumstances caused you and your team to adjust your initial plan to achieve the designated goal. Who determined the change in method/direction was needed? How was the altered plan determined? How did the team react to the shift?

Maintaining both your own motivation and your team's motivation over the course of a long-term project can be a challenge. Changing circumstances can be frustrating to team members. The eventual completion of the project may not reflect the originally planned journey; however, your team will arrive together and have weathered a uniquely shared experience. Leaders need to help team members continue to focus on the end goal, and the value their efforts contribute toward the project success.

High-Impact Motivators

Stakeholder management is one of the most important skills a twenty-first-century business leader can develop. Even the medical industry is putting more emphasis on interpersonal and patient-relationship skills as part of a physician's training program. Scott Magids and his research team explored *emotional motivators* that drive customer behavior. They discovered that successful retailers were able to "align themselves with emotions that drive their customers' most profitable behavior." When I read their list of the 10 high-impact motivators that affect customer value

Table 5.1 High-Impact Motivators That Significantly Affect Customer Value

I am inspired by a desire to	Brands can leverage this motivator by helping customers
Stand out from the crowd	Project a unique social identity; be seen as special
Have confidence in the future	Perceive the future as better than the past; have a positive mental picture of what is to come
Enjoy a sense of well-being	Feel that life measures up to expectations and that balance has been achieved; seek a stress-free state without conflicts or threats
Feel a sense of freedom	Act independently, without obligations or restrictions
Feel a sense of thrill	Experience visceral, overwhelming pleasure and excitement; participate in exciting, fun events
Feel a sense of belonging	Have an affiliation with people they relate to or aspire to be like; feel part of a group
Protect the environment	Sustain the belief that the environment is sacred; take action to improve their surroundings
Be the person I want to be	Fulfill a desire for ongoing self-improvement; live up to their ideal self-image
Feel secure	Believe that what they have today will be there tomorrow; pursue goals and dreams without worry
Succeed in life	Feel that they lead meaningful lives; find worth that goes beyond financial or socioeconomic measures

(Table 5.1), I saw many parallels that today's workplace managers and leaders face when trying to motivate reporting groups of people and project teams (Magids et al. 2015).

Advice and Motivation

Once people are motivated to put a stake in the ground and become active team members, they begin to build collegial relationships as they learn how to work together. Demonstrating their strengths and leaning into other's talents requires various *water testing* methods. People may feel they are "making it up as they go along," because there are no concrete blueprints or standard operating procedures to reference. Workplace pioneers involved with a new collaborative effort need to be comfortable with some level of uncertainty. They will look toward their managers, colleagues, and industry experts for advice. The encouragement, above and beyond the content, of these formal and informal advisory sessions has a strong impact on the seeker's continued motivation to pursue and explore.

A major role for collaborative team leaders is to provide support and guidance to help the teams stay on track. Have you ever gone to your team leader with a request for advice, but they did not really provide you the guidance you were hoping to receive? David Garvin and Joshua Margolis summarize some of the challenges between advice seekers and advice providers in their article, *The Art of Giving and Receiving Advice*. The basis for the article is the discovery that one of the most common sources of misunderstanding when someone seeks out guidance is the advice giver's role within the exchange.

As a leader, you will find yourself on both sides of being an advice seeker and an advice giver. All too often, people seeking advice presume that they already know the answer and are seeking validation. They then choose a like-minded advisor with similar experiences and viewpoints. These advisory reviews will produce a plan for which the weaknesses to the plan are reinforced and the strengths weaken. The result is akin to inbreeding. Similarly, advice seekers may not realize the breadth of advice they need and might overlook asking advice from someone who could offer valuable insights. Some of the most valuable advice an advisor can provide is the insight to help them find the best advisors who can provide critical breadth or depth for their current purpose.

Garvin and Margolis organize guidance into four categories: discrete advice, counsel, coaching, and mentoring. Table 5.2 summarizes the guidance role that the

Table 5.2 Advice Giver Roles Matrix

Type	Activities	Desired Outcomes
Discrete advice	Exploring options for a single decision	Recommendations in favor of or against specific options
Counsel	Providing guidance on how to approach a complex or unfamiliar situation	A framework or process for understanding and navigating the situation
Coaching	Enhancing skills, self-awareness, and self-management	Task proficiency; personal and professional development
Mentoring	Providing opportunities, guidance, and protection to aid career success	A relationship dedicated to building and sustaining personal effectiveness and to career advancement

advice giver plays on the basis of the type of information the seeker needs for that situation. Discrete advice is a query for specific information. More than likely, you are able to quickly respond to the seeker's question or you will be able to forward them on to someone who will be able to help them. When requesting counsel, the seeker is usually looking for advice to help them create a framework to make a decision or take action. As the person providing counsel, you should avoid providing a specific prescription. The Seeker is looking for suggestions for a way forward as opposed to a specific suggestion. During your conversation, they may realize that one path will be better for them from another, but it should evolve from their thought process not yours.

When seekers request coaching or mentoring type of advice, they are looking to develop an on-going relationship. These two types of guidance often are interchanged; however, they represent two very different types of relationships. A coach helps the seeker develop proficiency for a special skill, such as learning how to deliver a more effective presentation. Making a commitment to be a mentor to someone is much more about helping the seeker develop professional relationships for career advancement. Mentors will provide career and introduce the seeker to various people within their network for the purpose of gaining exposure.

Independent of which type of advice someone is seeking, an advisor has the opportunity/responsibility to provide a constructive framework that the seeker may not have considered. The advisor could be a sounding board to offer a few insightful questions for the purpose of clarity or help the seeker reshape their perspective by offering information that expands their breadth or depth of knowledge regarding the situation. If the seeker wants to walk through a tentative plan, the advisor should review their assumptions and offer hypothetical situations to plan for possible consequences (Garvin and Margolis 2015, pp. 61–71).

CONSIDER AND DELIBERATE

How would you respond if you asked someone for general guidance and the advisor directed a specific course of action with which you did not agree?

Describe a time when someone asked you a clarifying question that gave you some additional perspective. Were you grateful or grumpy during the interaction?

What standard clarifying questions do you ask people when you are trying to understand more about the question they are asking you?

How might you initiate a coaching session with someone who does not realize they need to improve a skill?

What information would you find beneficial from a mentor for your professional growth and development?

When people reach out to you to ask a specific or broader question, it is important to take a moment to assure that you are providing the advice or counsel being asked. Some days are hugely busy, and you may not be able to fully confirm that the information you provided was helpful. Don't be afraid to circle back to the person who asked your advice to verify that they have the information they need. If the seeker has what is needed, he or she will feel good that you respect them to be concerned. If you misunderstood what was needed, apologize that you could not provide your full attention and offer to discuss the issue now or at a scheduled time.

Different People Need Different Motivations

Motivational techniques are not just for your direct reports. Depending on the situation, you may be called on to motivate a variety of people including folks who are your peers and people who are senior to you. Official and unofficial leaders need to use all of the tools they have to make sure that all involved in the project are able to focus on the goal and avoid any time-wasting actions. Feedback, both nonverbal and verbal, is even more important for the motivational process among colleagues. Your peers may have opinions and suggestions that will provide value to you. More often, than not, you will need to adapt your motivational message for different audiences depending on their personalities and previous experiences.

The ritual Passover meal is a prime example of adapting your motivational message to different audiences. The story of Moses leading the Israelites out of Egypt where they had been enslaved for several generations to freedom on the other side of the Red Sea is told. Rituals involving symbolic foods and wine help participants to imagine that they are one of the slaves who are trying to escape to freedom and the lessons they learned along the way. The story and the progress of the holiday meal are orchestrated by a family member or friend who is encouraged to frame the story in different ways so that all will be able to understand.

The guidelines for this segment of the holiday meal instruct the person who is leading the Seder (ritual meal) to adapt their explanation of the Passover story for four types of children (learners): the wise one, the wicked one, the simple one, and the child who does not even know how to ask a question. The wise child requires the leader to explain the story in great detail and nuance. The wicked one has an attitude and considers gathering family and friends together to celebrate the achievement of freedom to be a total waste of time. For this child, the leader must work to help him/her understand their place in the story and how it relates to modern-day challenges. The simple one wants to learn but has difficulty comprehending complex explanations. For this child, the leader needs to provide a simply worded explanation. Moreover, for the child who does not even know how to ask a question, the leader is expected to introduce the story and help him/her to learn how participate.

This metaphor can be used in almost every change situation as you try to facilitate new skills training and motivate team members to work together to achieve a common goal. There is the group of people who love new challenges and want to know every detail. For these people, leaders may want to solicit their assistance to help smooth the transition. There are those people who will drag their feet and find the negative in anything that is new or different. Leaders will need to help these people understand the benefits that the change will provide to them, the team, and the entire organization. There are people who are excited but are nervous about significant change. For these people, leaders might want to move slowly—one step at a time. Moreover, for those, albeit few, who do not understand how the change will affect their work effort, leaders need to take a step back and explain how the tasks that these people perform will fit into the larger vision.

Mapping Motivation to Values

There are no simple answers when it comes to successfully motivating each individual working on a collaborative team. In addition to the normal diversity of personalities and cultural traits, project teams can consist of people representing different departments or partner organizations that support different workplace cultures.

Zachary Wong, an experienced project manager, shares a simple matrix to help people quickly map preferred motivation techniques for four general personality types in his book, *Personal Effectiveness in Project Management*. Table 5.3 provides a brief summary of Wong's comparison. It is important to remember that depending on the situation and their role, a person could easily take on any one of the four personas Wong identifies. This is a great tool to use when your team is forming and you are trying to make a relatively quick assessment of the structure of the team and the different motivational approaches you may need to take (Wong 2013).

Table 5.3 Personality/Motivational Technique Matrix

Personality	Characteristics	Motivations
Rational	Objective, analytical, technical, logical	Being challenged Feeling recognized Having autonomy
Idealist	Amiable, caring, collaborative, hopeful	Having a purpose Feeling respected and trusted Collaborating
Guardian	Organized, diligent, reliable, compliant	Getting the job done Feeling needed Feeling appreciated
Artisan	Creative, fun-loving, open, self-expressive	Having freedom to act Feeling unique and potent Having admiration

In these instances, experience is very helpful as you observe and sort the various motivational requirements needed by your team members. My only counsel is, do not explicitly identify and separate people according to their motivational profile. Use this knowledge as you speak with them one-on-one conversationally or during coaching situations. Your professional expectations for excellent work should not differ, only the context of your communication.

Essentially, as a team leader, you are a salesperson who is trying to sell a product such as an automobile. Different customers have different priorities. Your job is to figure out if the person in front of you is more concerned about safety, performance, style, or price. Once you identify their motivation to buy, you must shape your pitch accordingly. The goal to sell a car is the same for person walking into the showroom; the method you choose to use is slightly different. This is similar to Sitting Bull's strategy discussed in Chapter 1 when he addressed the specific strengths and interests of the individual Sioux chiefs to motivate them to align with him and his plan to defeat Custer.

An example of how you might shape your motivational strategy for different team members can be taken from Sitting Bull's playbook (Chapter 1) and use flattery as a means for motivation. During an organizational meeting, discuss team member's strengths or professional growth opportunities as you are making assignments. In the case of Rational Ruth, you might say, "Ruth, you did such a great job on the Configuration Management requirements on the last project, I would like you to take on …." For Idealistic Isaac who works best as part of a group, you might say, "Isaac, I appreciate you being able to uncover the essential feature in the procurement application. It would have been a disaster, if we had deleted it from the upgrade as we had originally planned. Would you take the lead coordinating with management to develop the end user training functions?" For Guardian Greg, you might say, "Greg, I'd really like you to monitor each team's progress to make sure we can stay on schedule. I appreciate your diplomatic abilities. I know I can count on you to establish trust with the team lead and you will be able to let me know when something unexpected is going to alter the schedule or budget by more than 5%." Lastly, you might approach Artisan Amy, "Amy, I'd like to do something a bit different with our weekly status meetings. Will you help me develop some thoughts that will take traditional checklists to a different level?"

CONSIDER AND DELIBERATE

Reflect on a time when a manager demotivated you. Was it the specific task assigned to you, or was it how the task was assigned? Please explain.

Describe a time when you were assigned to a relatively unexciting task or project. Did you plod along or did you find a way to visualize the larger picture? How did management relate to you during this project?

If you were managing people assigned to a less than exciting task or project, how might you motivate them to put forth an exemplary effort?

Many of us are self-motivated and take great pride in our individual work efforts. However, at the heart of exemplary teams who consistently perform above and beyond expectations is a leader who is adept using a variety of motivational techniques.

Summary

Focus and Motivation
- ■ Leaders must constantly monitor all available feedback channels, be attuned to trending changes, and enable adaptations to empower teams toward success.
- ■ The hierarchy of organizational needs and communication merges Abraham Maslow's hierarchy of needs model for individuals with an organization's communication requirements using the journalistic lens of Who, What, When, Where, and Why.

- Bringing people together in a workplace with different cultural and professional experiences requires leaders to use a variety of motivational techniques.
- Maintaining a team's motivation over the course of a long-term project requires a leader to adapt motivational techniques enabling team members to continue to focus on the end goal and the value their efforts contribute toward the project success.
- A major role for collaborative team leaders is to provide advice, support, and guidance to help teams stay on track.
- Recognize and encourage the unique strengths of team members.

Chapter 6

Growth Fueled by Uncertainty

Adaptation drives …
Failure helps …
Uncertainty provides …

Throughout history, many leaders have been hesitant to step forward to lead change. Sometimes it is because they are fearful as in the case of Steve Wozniak. He admitted not wanting to start a company because he was afraid. It took a concerted effort by Steve Jobs, friends, and his parents before he gained the confidence to leave steady employment and put his full energies into creating Apple Computer. Michelangelo refused the Pope's request to paint the ceiling of the Sistine Chapel for two years because he believed he did not have the talent needed to do justice to such an overwhelming commission (Grant 2016, pp. 12–13).

Sometimes, other people may see your potential and push you forward. Other times, you are the one who needs to look at a situation with fresh eyes to see what you can contribute. In addition, there are times when everyone else in the group steps back leaving you to rise to the occasion and figure it out. My grown daughter and I were chatting one evening when I asked her to share the first time she thought of herself as a leader. She reminded me of the evening when she was in the ninth grade and she and her 14 best friends went out to a restaurant before attending the high-school homecoming dance. Our family had two large cars and was volunteered as drivers. Her father, brother, and I had dinner in a different part of the restaurant. I noticed several of her friends were milling around and went to check on the rest of the girls. It turns out that each 14-year-old young woman had been given a $20 bill to purchase dinner; however, the restaurant did not offer them separate checks. One girl had ordered a steak. Two other girls had shared a salad. Even an experienced CPA would give the situation a big sigh. My daughter had stepped forward to figure out the bill and make appropriate change for her friends as well as managing the tip. Although she was happy to see me and visibly glad for my offer to help her wade through the final transactions, she had done the bulk of the work before I arrived at her table. She simply needed reassurance. Fourteen years later, she told me her friends continue to defer to her to divvy the check when they are together.

I was very proud of my daughter that night when she took a big breath and buckled down to work through the complicated dinner transaction. The onus lies with the parents (including me) who assumed our relatively level-headed, honor roll daughters who had purchased movie theater tickets, meals out with a few friends, and other sundry products would have no difficulties with the restaurant bill. None of us foresaw the complexity of a single check for 14 girls. The next year, when the group of young women all wanted to attend dinner and the homecoming dance together, they chose a restaurant that offered them a prix fixe menu with four entrée choices so that each of them knew exactly how much money to bring. The bill was so much easier to navigate because they were able to take the lessons learned and devised a solution that smoothed out the unforeseen complexities from the previous year.

Have you ever found yourself in a situation in which it turned out to be more complex than you or your manager had anticipated? As you took that deep breath and began to work through the various challenges, I bet at the end of the day, you learned more and gained confidence from having to handle the unexpected. This is where the lessons learning stage is so important. Given a similar situation, how can you mitigate the risks to prevent a similar situation from occurring?

Uncertainty fuels personal growth because it means that the challenge may never have been done before; or if it has, it has not been done within this type of environment or magnitude. Change happens because the status quo is no longer working. Very few people own a crystal ball that can foresee the future, and so, it falls to formal or informal leaders who take a deep breath and forge forward into the unknown. There are times when people feel they are making it up as they go along. However, if they have built a strong and trusted team, they will not be alone.

Survival by Adaptation

Complex change situations are often made more difficult due to a lack of leadership, practiced procedures, and effective planning. Many students of organizational complexity look toward nature, which is full of complex biological systems, to provide visual descriptions. One of my favorite examples is Stephen Johnson's description of the workings of an ant colony in his book, *Emergence*. I discovered that ant colonies' underground habitats are exactly the same anywhere in the world. This information is innate within the ants' DNA. Every ant environment contains the same set of chambers. The birthing area, the food storage warehouse, the communal space, and the funerary room all have the exact same dimensions and locations. This standardization is even more fantastic when coupled with the fact that there is no identified work supervisor. All work is performed methodically using scent sensations to communicate.

It may seem odd to bring up steadfast ant colonies in a chapter about change, but their existence is a perfect case for the concept of *survival by adaption*. While visiting a national park in Costa Rica, I observed a line of ants carrying food back to their underground storehouse. I suddenly felt very guilty for all those ant hills I had smashed as a child. How do ants survive those random external acts that disturb their systematic way of life? The answer is simple— they persevere. For example, if an established tunnel is suddenly impassible, the ant that encounters the challenge begins to clear an alternate path and leaves a predetermined scent marker to turn right or left for the ants that come after them. There is no established ant hierarchy to discuss the cause of the change or the various alternatives. The ant that encounters the impasse uses an innate capability to determine the necessary path and continues forward (Johnson et al. 2002).

The lesson is the same for the twenty-first century organizations to survive. The people who run them must learn to recognize changing situations and adapt products and services appropriately to survive increasingly complex environments. Essentially, change is becoming the status quo for many organizations.

If change is becoming a natural state of being, how do people adjust? Some people thrive on changes in their lives. Others prefer stability and the knowledge that everything is in its proper place. It is a matter of how people prefer to control their choices in alignment with their personal values and preferences. Imagine what happens when an assortment of people with a variety of experiences are brought together and told that management has made some well-thought out decisions that will affect their work environment. Management-speak will describe growth opportunities, changing roles, new skill requirements as work groups are reorganized, merged, and spun off.

Often it is not the specific change that is occurring; it is the uncertainty of not knowing how people will need to shift within the new environment that is so disconcerting. People may need to adapt their perception of their place within the changing organization structure. They need time and guidance to readjust the lens they use to align their skills knowledge, and perceived value. What behaviors will they need to adapt to succeed within the new environment? What latent capabilities do they need to hone that were not previously required?

CONSIDER AND DELIBERATE

Are you someone who enjoys change or do you prefer stable situations?

How do you feel when you find yourself in an opposite situation to your comfort level?

Do you prefer repetitive work assignments, or are you always the first to volunteer for a special assignment?

What is your favorite household task to perform? Why?

Being able to recognize your discomfort is the first step toward reconciling and acclimating to the change situation that is taking place within your current environment. The final two questions may seem out of place, but if part of our goal is self-knowledge, these two are telling. Most of my professional career, I have worked on large, complex, long-term projects. Even, if we could define a 'done' metric, it would change before we could achieve it. My favorite household task is cleaning bathrooms. Within a 15–20 min timeframe, I can see the shiny result of my efforts. There is a beginning, middle, and a definite end to the task. It is my therapy to counter balance the constantly changing nature of my work week.

Improvisation and Change

Without a doubt, change is down-right messy. It is difficult to plan every single detail, no matter how carefully people script the process. Change causes disruption. It may be a mild disorientation while people work through the 'oops, I forgot stage,' or it could be a total chaotic disruption to what incumbent staff thought was a perfectly fine way of doing things. When things get confusing, and a workable solution has yet to be realized, people will reference to Friedrich Nietzsche's quote, "That which does not kill us, makes us stronger."

Tim Harford addresses the potential that disorder can provide in his book, *Messy: The Power of Disorder to Transform our Lives*. He theorizes that the chaos of change pushes us to reach further into the deepest depths of our personal reserves and wrench out solutions that were never before considered. The enhanced creativity and increased resilience we acquired through this process would not have been realized had the transformation process been relatively tranquil (Harford 2016, p. 5).

In hindsight, the memories of collaborative teamwork, brainstorm sessions, and applied creativity are exhilarating. When the teams who worked through the mess were able to sit back and assess their accomplishment, they all breathed a great sigh of relief. The actuality of the change process that they had survived had been exhausting and yes, they were more creative and resilient than they had ever imagined. The significant portion of their success can be attributed to the use of theme and variation also known as improvisation. In just about every instance, people got down to the business of solution finding by trying to identify similar situations. They knew what they were asked to do had never been done before; but, they knew if they could find similarities, they could take the information forward to have a base from which they could start the solution development process and become a stronger team.

Artists use improvisation as a tool to inspire creativity and innovation. They take a simple basic theme that might be a few bars of music or a series of dance moves and create an epic work of art. They do come back to the basic theme throughout the piece and may take off in a totally different direction; however, they return to the theme, which can be referenced as their center or North Star to navigate their creative journey. Business leaders improvise often because there is not always time for an in-depth analysis. They have an inkling of what has worked in the past, superimpose the current environmental situation, and improvise a pathway from which to start the solution process. Essentially, improvisation is on-the-fly creativity. Sometimes it is successful and sometimes it is not.

Miles Davis is famous for the jazz album, *Kind of Blue*. This album that is revered among jazz aficionados was produced using a total improvisational experience. Davis had a vague idea of what he wanted, though he was dependent on the talent and experience of the musicians to help develop his basic melody sketches. Mistakes were incorporated into the recording. The finished product that was recorded over two sessions in less than nine hours was not exactly what Davis had intended; but it became a masterpiece based on how the energies combined. Unlike *Kind of Blue,* the Beatles' *Sgt. Pepper's Lonely Hearts Club Band* was recorded and produced in a series of sessions totaling 700 hours. Although each album was developed, recorded, and produced using different methodologies, both are masterful efforts (Harford 2016, pp. 95–98). It comes down to the amount of time available and the level of control desired by the artist. If you have a minimal amount of time and trust your team, improvisation may be an excellent option. If you are a perfectionist with few time constraints, you will prefer a creation scenario that allows you the most control over the finished product.

Business-based improvisation happens when there is little time for extensive planning. Some people even prefer improvisation because they do not necessarily have the patience to work through a methodical preparation process. On the other hand, people who prefer controlling all elements that surround a situation will be very frustrated. Improvisation, as demonstrated by Miles Davis and other artists, supports limited, if any, centralized control. This is what enables the creativity of the process. The trade-offs and risk factors to consider when deciding if an improvisational methodology will be beneficial are the schedule, budget, and flexibility of the requirements.

Like any skill, the more people practice, the stronger their improvisational skills will become. Tim Harford summarizes the traits needed to learn how to successfully improvise into the following four categories:

- Practice, reflection, attention
- Accept the messiness/confusion of the situation
- Listen intently
- Let go of control, balance risk

Practice is being able to pay attention in the moment and to reflect on the direction a consequence will take you. Have you ever wondered how quickly improved comedy groups move from gathering random words from the audience to a fairly funny, somewhat complicated skit? Hours of practice helped them to develop trust and a level of comfort with each other that allows them to let go of control and adapt to the flow of the scene. Even if a comedian preferred a scene to take a different direction, they must follow the riff that their comrade began. If they are not listening intently, they may miss a queue to respond or an opportunity to make an impact and get a big laugh (Harford 2016, pp. 109–111).

CONSIDER AND DELIBERATE

Briefly describe a difficult situation that made you stronger and more resilient.

What skills did you depend on to help you survive and thrive on the challenge?

Reflect back to a situation in which you found yourself put on the spot. What were you thinking in the split seconds before you had to improvise an answer?

What steps would you take to prepare your team to improvise a solution to a complicated situation?

The first action to take when faced with a complicated situation is to take at least two very deep breaths. The purpose of these two deep breaths is to get oxygen to the brain so you can think more clearly and to give yourself a bit of calm from which you can focus your thoughts before you begin to improvise a solution to address the challenge of the moment.

Leading through the Paradoxes of Change

Whether working within an improvisational mode or developing well-thought out plans, leaders within changing organizations need to guide and align a shifting multilayered mosaic. There will be times when leaders may find themselves managing a variety of inconsistencies as they try to balance the demands to achieve both short- and long-term goals. Wendy K. Smith and her research team reference these multidimensional practices as Paradoxical Leadership. By using improvisational strategies, leaders are able to create an organizational mindset that enables team players to create an abundance of possibilities from the same set of resources that traditional leaders label as scarce.

Instead of making trade-offs that provide multiple groups thinner and thinner slices of the resource pie, paradoxical leaders look to grow the pie exponentially through collaborative strategies and resource reframing for flexible outcomes; and thus, the paradox—How do some people given the same resources and similar scenarios create so much more value? The answer is that the successful leaders recognize and address each element/layer of the dynamic organization mosaic that will incorporate

- Multiple values
- Variety of contributions
- Unique differentiators

Table 6.1 Paradoxical Leadership Matrix

"Either/Or" Leaders	*"Both/And" Leaders*
Acts consistently	Acts consistently inconsistent
Make strategic choices	Simultaneously engages in conflicting strategies
Sets a clear agenda	Searches for opportunities to grow resources
Makes allocation trade-offs	Sets flexible time frames
Adopts consistent identity across organization	Embraces multiple strategies and identities
Promotes best practices	Tolerates uncertainty
Keeps it simple	Learns from failure

They embrace the change, the chaos, the mess, and open themselves to opportunity and possibility. They will leverage experimentation, critical feedback, and continuous tuning as they guide their teams toward a solution. Their secret weapon is a variation on the improvisational game, *Yes, and*. When improv groups play this game, one person starts with a statement such as, "Let's pack a picnic and go to the beach." It does not matter, if the second person would much prefer the mountains, they must agree and add another element to the story. The second person might say, "Yes, and we'll stop on the way to buy rocket launchers to use after dark." The story builds and expands.

Wendy Smith and her team categorize traditional leadership approaches as using *Either/Or* strategies and the paradoxical leadership approach as using *Both/And*. Table 6.1 compares the various behaviors of these two types of leaders that can be summarized by consistently performing leaders who mostly focus on allocation trade-offs and leaders who explore opportunities to enhance the value of current resources (Smith et al. 2016, pp. 63–70).

CONSIDER AND DELIBERATE

Identify two or three departments that impact your work projects. Sketch the organizational linkages.

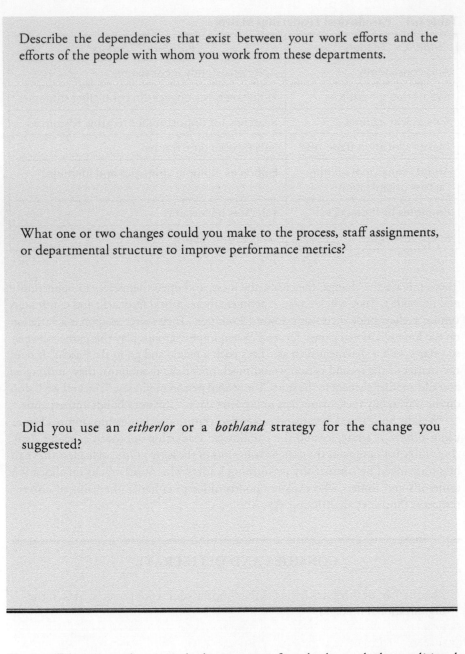

Describe the dependencies that exist between your work efforts and the efforts of the people with whom you work from these departments.

What one or two changes could you make to the process, staff assignments, or departmental structure to improve performance metrics?

Did you use an *either/or* or a *both/and* strategy for the change you suggested?

There will be times when it is the best strategy for a leader to deploy traditional leadership skills to drive strong team performance. However, when you are juggling multiple priorities, organizational complexities, and the multilayered effects that occur with change, you may want to consider experimenting with a few paradoxical leadership strategies to help your team work through confusion and uncertainty.

Change, Transitions, and Uncertainty

When several organizational groups or teams merge, the complexity of the ensuing uncertainty exacerbates. What ties this group of individuals together? How are multiple contributions linked together? Will various groups overlap? People may suddenly split their time between two managers. How will work priorities be negotiated across workgroups?

These alterations are not necessarily negative and will soon become the new normal; and so, in addition to trying to realign their professional pursuits, people start to wonder how established workplace social norms will change. What will happen to the norms and traditions that surround daily coffee klatches, lunch breaks, and informal watercooler conversations? How will they shift? The effect on their informal social group may have a greater impact on them than the official organizational change announcement.

Working through change, team members, as well as leaders need to gain confidence as they address the accompanying uncertainties. Stephen Heidari-Robinson and Suzanne Heywood cited a McKinsey study that surveyed 1800 executives regarding the most common reorganizational pitfalls in their article, *Getting Reorgs Right*. Employees resisting change was cited as the number one reason why reorganizations failed. Why do people find change to be so difficult? The author's research indicates that two-thirds of company reorganizations provide some business improvement. In light of a positive potential, why are employees resistant to change that can deliver positive results? The short answer most often is how the mission, goals, and process that drive the reorganization are communicated. Other reasons can be attributed to previous change experiences in which they carry forward memories from prior failed experiences cited by the McKinnsey results which include

- Insufficient people, time, and money dedicated to the reorganization effort
- Individual productivity declines because of distractions from day-to-day operations
- Organizational chart changes do not filter down to the functional level
- Unforeseen consequences (such as expanded IT requirements) derail the implementation plans

Heidari-Robinson and Heywood recommend that organizations take a systematic approach when planning a reorganization process has a significantly greater chance for success. Figure 6.1 illustrates the flow of the suggested process to best realize successful results. The first item to remember is that a reorganization process is not just about moving people around; it also is making a change to the business. Executives need to construct a profit/loss statement for the endeavor. What are the costs? What are the benefits? What are opportunities? Moreover, what are the risks? The second step is to review the strengths and weaknesses of your resources.

Figure 6.1 Systematic reorganization approach double-loop learning model.

Where are the gaps? What level of investment is required? Will you need to hire additional people? What type of internal training will be needed? The third step is to determine how the changes will be implemented. Will the changes be deployed from the top down or functionally from the bottom up? Will you have a beta group as an initial test group from which to learn and adapt the plan before a full out implementation? The fourth step is to clarify the implementation stages to make sure that resources are available when needed. The fifth and final step is to monitor the change process carefully so that mistakes can be corrected, any adjustments verified, and the reorganization plan updated to clarify the learned changes (Heidari-Robinson and Heywood 2016, pp. 85–89).

Very often, reorganizations and other types of changes happen because someone suggests a good idea and it expands into a reality before the details can be thought through, much less a plan developed. A Nigerian folktale explaining why the sun and moon live in the sky is an example of how the consequences from an idea that is not well-thought out can have a long-term impact.

Many years ago, Sun and Water were great friends. They both lived on the earth together, and Sun would visit the Water at his home; however, Water did not visit Sun. One day, Sun asked Water why he never visited the Sun's house. Water responded he needed to travel with all of his people and Sun's house was not big enough to hold them all. Sun went back to the house he shared with his wife, Moon, and began to build a huge compound for Water and his people to visit. When he finished building, he invited Water to come for a visit. The next day, Water arrived and began to flow into the compound that Sun had built. Soon, Water and all of the fish and sea animals began to overflow the compound. Water asked the Sun, if it was still safe for him and his wife. Sun, being very polite, answered, "Of course, you and your friends are most welcome." Water and his friends continued flowing. Soon the water level had reached the top of Sun's head. He stood on a chair to remain safe. Water called out again to determine if Sun and Moon were safe. They were very polite people and wanted to welcome their friend and responded, "Of course, you are most welcome." The Water kept flowing until Sun and Moon were perched on the top of the roof. They felt it would be impolite to stop their friend and his friends from entering their home and would not call a stop to the flowing water. Finally, they were forced into the sky, where they remained ever since (Dayrell 1910).

Sun had the very best of intentions, but he did not realize the immenseness of the sea. Water made checks regarding the well-being of Sun and Moon, but they were too polite to withdraw their original invitation. This is exactly what can happen during reorganization. The initial suggestion sounds like a great idea. When unforeseen circumstances occur, no one pays attention to how the change might affect the original plan. The negative reality happens, and people spend way too much time pointing fingers of blame at each other for not having foreseen what could happen.

CONSIDER AND DELIBERATE

Reflect back to a time when an off-the-cuff suggestion you made to your friends or within a professional brainstorming was determined to be the *perfect idea*. What were your initial feelings?

Did you try to discourage the group from carrying the idea forward?

What was the final outcome?

What sort of memories do you have regarding that event today?

Brainstorming sessions are excellent for helping people think outside of the proverbial box; however, no idea should be taken forward until it is properly vetted. My 'wish I had stopped it story' story happened during a meeting that was supposed to determine the next new technology around which to create a curriculum. I threw out an esoteric concept expecting everyone to shut it down. They surprised me by unanimously voting it forward. It was a disaster. We did not put much effort into the pilot course for which not a single person registered. The time and money lost from the failed effort was minimal. The lesson I learned regarding the necessity of thinking through a 'really' good idea before moving into the deployment phase was invaluable.

Uncertainty and Adaptation

Reorganization is just one of the types of changes businesses encounter throughout their lifecycle. Similar to natural organisms, businesses whether for profit or nonprofit must adapt to survive changing environments. Martin Reeves, a senior partner at the Boston Consulting Group, Simon Levin, a Professor of Biology at Princeton University, and Daichi Ueda, an associate at the Boston Consulting Group, merged their disciplines to explore how the twenty-first century organizations can prepare to address unpredictable environmental risks and ward off organizational threats that can cause their demise in the article *The Biology of Corporate Survival*. Their research acknowledges three complex trends that increase an organization's uncertainty and risk:

1. Unpredictable business environments
2. Shorter, fast-paced technology change, and integration cycles
3. Global partner interdependencies for which local businesses have exposure with minimal control

The authors' concern is based on the complexity of today's business environment and the inability of organizations to recognize and adapt to shifting circumstances. Their recommendation to organizations trying to prepare a resilient foundation is to expect surprise, but reduce uncertainty.

Paying attention to the full environment allows leaders to recognize details that could signal change. They suggest a strategy of employing a heterogeneous staff from diverse backgrounds whose depth of experience enables the team to visualize possible outcomes and prepare actions to mitigate negative consequences. Regular feedback loops and modular reporting structures will enable organizations to adapt and respond as needed to unexpected changes.

One example, they discuss, is The Montreal Protocol that is an international treaty designed to protect the ozone layer. Scientists from around the world shared hard data that analyzed the impact of fluorocarbons and the depleting ozone layer on human health. They lobbied international governments and businesses to take action, and the result is large-scale international cooperation for which the results demonstrate a measurable restorative impact on the atmosphere. As the initial accord was passed in 1987, The Montreal Protocol has undergone eight revisions to address new information realized through the monitoring process and continues to be considered one of the most successful international agreements.

The lessons the business sector can take from this is to actively monitor your market environments to become aware of small signals that could negatively impact your organization if they were to gain a strong position. Well-established market leaders make a mistake when they rest on the laurels that took them to their market leader position. Ignoring your competitors' actions and especially the innovations from small new market entrants is a signal that you have stopped adapting and may not survive as a company (Reeves et al. 2016, p. 55).

Adaptation requires leaders to recognize that they, their staff, and even the functional structure of their organization can be improved. In fact, uncertainty can be seen as an opportunity to give your organization a diagnostic once over and reframe some long-held ideas as the first steps toward innovation and growth.

Reframing Uncertainty

The first step toward addressing uncertain situations is to try to find a common understanding. This may take the form of viewing the situation from a different perspective that many call reframing. The benefit of viewing the situation from a different angle is that you might discover additional information that you have not yet considered. This information may help you identify some potential solutions or options for action that can be explored, evaluated, and drive well-informed decisions.

Reframing is a corollary to the adage, "if your only tool is a hammer, every problem requires a nail." This is the strength of collaborative learning. Discussing the circumstances and resolution possibilities among several people provides the opportunity to explore a situation from a variety of perspectives. Someone might bring up a possible consequence no one had considered. A solution that is cheaper, faster and provides improved quality might just be the outcome from these creative conversations.

In addition, sometimes the solution is so simple; the reframing process might help one to cut away complicated operations that evolved over months or years. A fresh set of eyes often is able to see the obvious that no one was able to recognize. The story about the young boy who told the emperor he was wearing no clothes illustrates this point. A self-proclaimed haute-couture tailor traveled through the kingdom and asked to see the emperor so he could make him an outfit of the most beautiful fabrics. The emperor was a bit vain and invited the tailor to stay at the palace while this beautiful outfit was made. The tailor secreted himself away and would let no one see the outfit until it was complete. On the day of the grand unveiling, the tailor carried what looked like an empty hanger into the emperor's chambers. The tailor immediately asked the emperor what he thought of this beautiful outfit made of gold and silver fabrics. Not wanting to appear stupid, he went along with the tailor. The tailor personally dressed him using elaborate pantomime. The emperor's closest advisors were called in to see this most elaborate outfit. The advisors immediately began to praise the beautiful fabrics that the tailor was describing to them because they did not want to appear stupid and unworldly.

Someone suggested the emperor should walk about his kingdom so all of citizens of his realm could admire his new finery. As the emperor walked through the main square, a young boy called out, "The emperor is not wearing any clothes!" Guards immediately snagged the young man. The emperor called out for him to be let go because he was telling the truth—he was not wearing anything but his undergarments. He had been a vain and foolish old man who had been duped into believing something that was not true. He called for the fraudulent tailor who was nowhere to be found. He chastised his advisors for telling him what they thought he wanted to hear and praised the young boy for his courage and honesty.

The young boy was not a highly paid consultant. He simply stated what he saw. The complicated description spun by the tailor and the court sophisticates of what they thought they were supposed to see was nothing but a fraud. This simple tale is a parable for what can happen when complex, multifaceted organizations are confronted with what appears to be complicated decisions. It is important to promote simple conversations that assure understanding of the root cause of the problem and not limit efforts to resolving the symptoms.

Active listening is a crucial behavioral component for successful leaders. It involves both questioning and listening. How many times are you in a meeting or at a social function where you can honestly say that you are providing the other person your absolute full attention? An active listener provides the other person(s) their full attention and demonstrates this by asking probing questions and listening to the entire answer before commenting. The goal of the conversational interchange is to reach a full understanding of the event, process, or challenge at hand. How often do people *assume away* complexity? They take information at face value without pausing to see if there might be some sort of challenge lurking in the shadows.

There is a story that describes a situation in which people ignored subtle symptoms that did not seem serious, at first. Steve, the factory supervisor, sees a puddle on the floor and calls for Joe to mop it up. Later that afternoon, he sees the puddle again and calls out to Joe, "I told you to mop up that mess, why didn't you do it?" Joe mumbles he did already and shuffles off to get the mop. Not once, does Steve question the source of the puddle. The next morning, the workers arrive to find the entire floor flooded. In researching the cause of the flood, it was discovered that the prior weekend a windstorm had knocked over a large structure that cracked the roof when it fell. The puddle was formed by water leaking through the roof. A major rainstorm had come through during the night. The tiles could no longer support the weight of the water as it gushed into the ceiling enclosure during the storm.

If Steve had stopped to research the source of the recurring puddle instead of demanding a quick fix to the symptom, the leaking roof may have been diagnosed and an expensive accident avoided. When dealing with multiple layers that support complex systems, people assume obvious symptoms as the actual cause for the challenge at hand and ignore the possibility that the problem could be a consequence of a deeper predicament.

Leaders who actively question and listen gain power because they know they do not need to do everything themselves. They gather people who have a variety of viewpoints and expertise to resolve complex situations. Examining a problem from several angles offers a better chance that the actual root cause can be discovered and resolved. Leaders demonstrate their strength when they are able to express confidence in the different working groups' success and align the separate efforts to achieve the organizational vision.

CONSIDER AND DELIBERATE

Has a manager or another influential person in your life dominated the conversation, assuming their situation was exactly the same as the one you were experiencing?

What was your reaction? How did the scene play out? Given another opportunity, what do you wish had happened?

Think about the managers, leaders, or colleagues who impressed you. How did they help you work toward complex solutions?

Many times the solution reveals itself when people take the time to ask questions as part of the process to identify the root cause of a problem. People often waste much effort and time-solving individual symptoms without stopping to consider how the various issues could be related. Albert Einstein is quoted, "If I were given one hour to save the planet, I would spend 59 minutes defining the problem and one minute resolving it."

Double-Loop Learning

Chris Argyris and Donald Schon, researchers in action-oriented organizational learning, developed a double-loop learning Model to help people address uncertainty and resolve complex problems. Double-loop learning as illustrated by Figure 6.2 is a series of iterative, interconnecting decision stage loops. The basic premise of double-loop learning is that each time a problem situation is viewed from a varied perspective, or a different method is tried, we learn a little bit more about the problem and the eventual solution. The five-phase learning process defined as Intention, Evaluation, Feedback, Decision, and Action also is

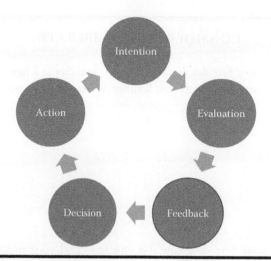

Figure 6.2   Double-loop learning model.

designed as individual iterative functions. For example, when contemplating how to make a significant change within your organization, it is important to carefully define the intention. What is the purpose or ultimate goal of the new program?

Moving forward into the evaluation stage, review the current environment and identify the various methods to create the new program. Identify the financial and procedural options/challenges. During the feedback stage, identify people who will be affected by the change and people who may have already worked through a similar process. Add their feedback to the information you gathered during the evaluation period. Have you not considered an important planning element? Are your priorities aligned properly? Are you asking the right questions and thinking about the right things? (Morgan 2006).

Once you have reviewed the feedback and begun to hone your plan, select a pilot or beta project as an initial test for the change action. If the trial action is successful, then you can begin to move toward the final decision and deployment phase. This model is relatively flexible and can be applied in almost every situation. For example, you may discover during the evaluation phase that the budget required to fully implement your intended change is much too expensive. You do not necessarily need to give up your mission. Your initial effort was not wrong; but, if you have budget limitations, the effort needs to be scaled back.

As you move from one phase to another, it is important to review the plans to assure that the activity/project continues to align with the organization's mission and that schedules and staffing resources are coordinated.

CONSIDER AND DELIBERATE

Remember back to a *first day*. It could be your first day at a new job or joining a new project team. What were you feeling? What was your perception of the environment?

At the end of your first day, how did you feel? Did your perception of the environment change?

What happened between the morning and the evening of your first day? What did you learn?

What about your last day at that job or project? (If you selected a current project, describe your current thoughts.) What were you feeling? What was the perception of the environment?

What occurred between your first and last day that shifted your feelings?

Most first days involve a great many challenges and unknowns that require navigation. Apprehension and excitement are normal feelings for first days. By the end of the first day, some people have a well mapped–out course; but, more often than not, people leave feeling a bit overwhelmed by the procedures and acronyms that come so naturally to colleagues at the new organization. Experienced leaders advise new leaders coming into an organization to embrace the uncertainty initially. They recommend that people watch and observe the culture, the processes, and how people work, before making any changes. Once you understand the organization and the people within it, you will make wiser suggestions that should be received more readily than if you immediately began making slash and burn changes.

Last days also can be difficult for a very different reason. No matter how exciting your next position may be, there is much comfort in being familiar with an organization's rhythms. The procedures and acronyms that were so foreign on your first day are second nature to you. Somewhere between your first and your last day, you experienced an individual learning curve. During change situations, both individuals and organizations must work through learning curves that are not always synchronized. Individual learning can be just as complex as organizational learning. Not all people within an organization learn in the same way and at the same pace. Leaders need to pay attention to the pace of the various learning rhythms. If the individual and organizational learning curves get too far out of synch, leaders need to step in to assure everyone, even in the midst of uncertainty, maintains the same trajectory alignment as they work toward their goals.

Risk, Uncertainty, and Planning Opportunities

There is always an element of risk in even the most simplistic of management decisions. What is the possibility that the project/event will be successful? Or, conversely, what is the possibility that it will fail? The role of a leader is to address risk directly head-on. Leaders spend the bulk of their time managing opportunities and expectations. They must forecast impending obstacles, identify the likelihood and significance of the impact, and mitigate any negative possibilities to ensure a positive outcome.

When you face a change situation, uncertainty adds an additional risk element to the decision-making process. Traditional risk management addresses the impact a proposed change will have on organizational resources and procedures. Risk/benefit analysis is fairly similar whether you are reviewing a simple solution for a predictable change outcome or a project that involves many moving parts and carries a significant amount of uncertainty. The challenge to developing and deploying a solution for a complex conundrum is the necessity for people to address the range of potential solutions from a multidimensional lens. When the problem resolution process is based on multiple viewpoints, the final solution will be more robust with a greater potential for achievement as it addresses a total organizational perspective. Involving representatives from the entire organization in different aspects of the change process will also help one to assure a successful change transition. Figure 6.3 illustrates the risk-analysis procedures applied to predictable change situations and the additional layers of analysis driven by multiple lenses that overlay the basic process as people strive to create innovative solutions in an effort to minimize uncertainty.

Every individual or organizational situation is different, in some way. Each time you are confronted by an 'opportunity,' whether positive or negative, ask the following questions:

- Is this a risk that needs to be addressed?
- What is the driving basis of this 'opportunity'?
- How will this affect the current organizational culture?
- Who needs to be part of the discussion and planning?
- What message needs to be communicated? When? And to whom?

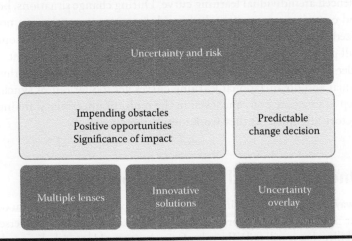

Figure 6.3   Unpredictable and predictable risk matrix.

Your answer to these questions will determine whether the situation is routine with a predictable outcome, or if you will need to use multiple lenses to develop an innovative solution.

The first step toward risk assessment is to determine the significance of the 'opportunity' and how it will affect the goal being pursued. Taking the time for this evaluation enables leaders to keep people focused on what is important and set aside what is not important. When a suggestion is presented as a positive benefit, the evaluation criteria should incorporate how the activity aligns with the organization mission (benefits) and the people and financial resource requirements to pursue the activity (costs). It is important to consider what is motivating the person who is bringing this wonderful opportunity to the table. How will each party benefit?

When the risk introduces a problem, it is important to gather as much information as quickly as possible to understand the significance of the negative impact. How badly will the planned activity be impeded? Is it a minor issue that is a distraction or is the situation, one which could change the direction of the project?

Once you have the facts and can identify the root cause and not just the symptoms, you and your team will be able to evaluate the situation rationally and then will be able to identify several possible strategies to resolve the situation.

Vijay Govindarajan, a professor at Dartmouth's Tuck School of Business, refers to these risk-evaluation activities as Planned Opportunism. His theory encourages people to mitigate the discomfort and worry that stem from uncertainty by preparing for those challenges that can be predicted. As people are able to exert some level of control in the face of uncertainty, they will not be as disturbed by those situations for which they do not have an immediate resolution. They are able to maintain a productive level of calm because the action they took freed them from the terror of passivity or no action. Teams are able to pursue possible resolutions with the knowledge that they are being as proactive as possible. The power of this technique to take control of uncertainty is the confidence people gain as they begin to develop their own change culture (Govindarajan 2016, p. 61).

Collaborative Resolution

The diversity of the team participating in problem resolution discussions and planning has a direct correlation with the success of the outcome of the decision solution. Everyone brings their personal experience and a certain amount of bias to every dialogue. You want to include representatives from all of the organizations who could be affected by the problem or the solution. Will the scope, schedule, or allocated budget need to change? What are the standard operating procedures?

Communications regarding the problem/solution should not wait until the implementation process is underway. The timing of communications can be as important as the message you are sending out. You do not want the rumor mill to be anyone's information source. An established set of procedures that can automatically be deployed when an unexpected situation occurs will help you to maintain control of the message context and format that people both inside and outside of your organization receive.

Have you ever shared good news with someone sworn to secrecy and heard it coming back at you on the grapevine? Had the meaning of the message changed? Even, if it was a positive opportunity, how much damage control was required? Could any of the energies required to 'make things right' been avoided if you had stopped for a moment to consider the risks?

CONSIDER AND DELIBERATE

Briefly outline the circumstances surrounding an unexpected 'opportunity' you encountered over the past three months.

What criteria did you use to evaluate the situation?

How would this 'opportunity' affect the originally planned outcome?

Who did you consult to help you determine the best solution? What is their role? Did you follow their advice? Why, or why not?

When and how did you share the 'opportunity' and the various resolution stages with interested parties?

Risk assessment, mitigation, and communication can be complicated even for relatively straightforward situations. The complexity of these procedures exponentially expands when multiple departments and organizations need to be brought into risk prevention and resolution discussions. Who, when, and how become significant details that require advanced attention.

In summary, change does not occur in a vacuum. The consequences from one action can create unanticipated consequences in other areas. Leaders must constantly be alert to positive or negative 'opportunities,' address them as they arise, and find the best people available to anticipate an appropriate resolution.

Summary

Growth Fueled by Uncertainty

- Organizations must learn to recognize changing situations and adapt products and services appropriately to survive within increasingly complex environments.
- People's discomfort with change is often more about the uncertainty of not knowing how the story will end than the actual change that is being deployed.
- Collaborative consultation and improvisation can bring creativity to the solution and avoid missteps.
- Effective change leaders incorporate multiple values, consult diverse points of view, and identify unique differentiators.
- The primary reason why reorganization efforts fail is due to employees' resistance to change. Communications aligning the organization mission, goals, and processes with the change effort can help one to gain employee support.
- Reorganizations change the business and should not be initiated without developing a well-thought out plan identifying all costs and benefits.
- Uncertainty is an opportunity to review and reframe long-held views.
- Examining a problem from several angles offers a better chance to discover and resolve the actual root cause.

Chapter 7

Super Powers for Leaders

Vulnerability helps leaders …
Mindless and mindful people view …
Emotional Intelligence maximizes …

There is a series of Laotian folk tales that center on a clever young monk named Xieng Mieng. One story is told in which the King complains he has no appetite. No one is able to produce a delicacy that will entice his taste buds. None of his advisors or wise men could produce a cure for the King to be hungry at meal times. In desperation, he granted Xieng Mieng an audience. The young monk told the King he had a perfect plan to return his appetite, if the King would follow his directions exactly. The King agreed and paid Xieng Mieng a handsome fee in advance of his services.

Xieng Mieng told the King that he had a special root from which he would make a soup; however, the King had to promise not to eat a single morsel for the entire day. If he ate anything, the root's curative properties would be negated. The next day the King followed Xieng Mieng's directions and ate nothing. Toward sundown, the King was cranky and wanted Xieng Mieng and his magic root soup. Xieng Mieng arrived without the root and no soup. The King roared, "I am starving!" Xieng Mieng smiled and told the King, "See, your appetite has returned." When the king realizes the simplicity of the solution, he believes he has been tricked and rages threatening retribution against Xieng Mieng. The young monk uses his cleverness to escape these threats, only to return the next time the King finds himself in an unsolvable situation (Halpern 1957).

Nor Sanavongsay, a San Francisco-based writer, explains, "Xieng Mieng tales are particularly beloved by the Lao because of the message that one doesn't have to be famous or wealthy to be successful in life. Anyone who's willing to be smart can find ways to laugh and prosper, whether in ancient Laos or America." Xieng Mieng's super power was his clever common sense that eluded all of the King's advisors. We all have innate super power abilities that will help us to resolve complicated dilemmas, if we allow them to rise to the surface. A strong leader is someone who is able to glean the essential core of a problem and work with his/her team to apply appropriate measures to resolve it.

CONSIDER AND DELIBERATE

What single leadership super power would you wish to possess?

Describe a scenario for which your super power would help you and your team remedy a difficult situation.

It would be so very simple if we could swallow a leadership pill, fall asleep, and wake up with all of the leadership super powers at our disposal. Reality is much tougher. After much research, practice, and observation, I have uncovered four nontangible super powers that successful twenty-first century leaders probably already possess without realizing the power they hold. These leadership super powers are vulnerability, educated ignorance, mindfulness, and emotional intelligence (EI).

Throughout history, leaders possessed and practiced these attributes; however, they would not necessarily admit to what many considered weaknesses. It is only

Table 7.1 Project Management Super Powers Matrix

Leadership Super Powers	PMI Talent Triangle Elements		
	Technical Project Management	Business Management	Leadership
Vulnerability	Risk management	Navigate uncertainty	Strategic humility
Educated Ignorance	Requirements	Scope	Multiple perspectives
Mindfulness	Attune	Acclimate	Adapt
Emotional Intelligence	Decision-making	Stakeholder management	Stress management

through recent scholarly research that these intangible characteristics have been measured and documented as part of successful performance scenarios. Reflecting back to Nor Sanavongsay's quote, it is how we define *smart* and thus the value we put on the characteristics and traits. Will we define *smart* as intelligence or a sharp stinging pain? Often, leaders will feel a sharp stinging pain while they acquire new intelligence, reinforcing the need for multiple perspectives.

When I teach project management classes, I always seem to get at least one hard-core person who believes leadership theories have no place within their *hard skill* project team. By the end of the class, I usually am able to persuade them to at least grudgingly agree that successful project management focuses on *the people*. Table 7.1 offers a matrix that demonstrates the overlap of Project Management Institute's (PMI's) Talent Triangle professional development elements and the four leadership super powers. The intersection points identify the work actions that project leaders perform many times over the course of the project life cycle. The rest of this chapter explores the foundation and application of the four leadership super powers represented by this overview.

Vulnerability and Leadership

When a crisis occurs, people look toward their leaders to be calm and to tell them what to do to survive the situation. All cultures have their version of the British *stiff upper lip* or *keep calm and carry on*. When my friends and I were learning to be young managers, our mantra was *don't let them see you cry* because we did not want to appear weak or out of control. Throughout history, leaders are seen as strong founts of strength and knowledge. Was I surprised when I was promoted into a leadership position and realized I could not possibly have all of the answers and found myself trying hard to look confident as I improvised solutions. I cautiously began asking people who I respected for advice. This is when I learned the greatest secret that leaders do not want to reveal is that they do not have all of the answers. They did not want to look weak and out of control.

A few years ago, I read Brené Brown's book *Daring Greatly*. Her explanation of how being vulnerable can transform leadership practices literally reached out from the pages, grabbed me by my collar, and made me take notice. She discovered the most common fears among C-level executives were being able to clarify a purpose, remain relevant, and maintain connections. Instead of exposing weakness, her research proved that by recognizing their fears, leaders can use this knowledge as a catalyst to promote creativity, innovation, and learning. Once you've identified the challenge—whether it is lack of information or resources—you are free to imagine what is possible. In fact, the result of refusing to recognize their vulnerabilities will cause leaders and the people on their teams to make excuses, dodge questions, and falsely blame others (Brown 2012, p. 15; p. 207).

Educated Ignorance

Jesuit priests, Father George Coyne, the former director of the Vatican observatory, and Brother Guy Consolmagno, the current director of the Vatican observatory, discussed the parallels of faith and science with Krista Tippett during a radio interview. They explained the parallels that exploration plays in both their religious and scientific vocations because both require them to be seekers in an environment of uncertainty. It is the awareness of what they do not know—their *Educated Ignorance*—that drives them to continue their spiritual and scientific inquiries. These thoughts are not unique to the Jesuits. Socrates spoke of educated ignorance when he said, "I'm wiser than everyone else because I know what I don't know." Brother Guy explains his ability to balance science and spirituality with a reference to novelist and writer Anne Lamott who wrote, "The opposite of faith is not doubt. The opposite of faith is certainty." in her book, *Plan B: Further thoughts on Faith* (Tippett 2016).

Leaders often cannot be certain of an outcome. Educated ignorance enables us to say it is okay not to have all the answers. It is okay to admit we need more information. It gives us a plan, at least a starting point, to move forward toward expanding our knowledge base to be able to make a decision by asking questions and listening very closely to the answers.

Past experience and newly identified data points are what leaders use to guide their teams to accomplish the mission. We become explorers in that we set a reference point, similar to the North Star, from which we can look forward toward new methods to overcome an obstacle that is impeding our progress toward our goal. This is when we rely on our risk management skills to analyze what we know to be true balanced against the possibility of outcomes that may or may not happen. Leaders are empowered and not inhibited by a lack of knowledge. The journey of exploration can be energizing (albeit sometimes exhausting) for both the leader and their team.

Mindful Leaders

Ellen Langer, a Harvard-based Social Psychologist, was curious about the traits that differentiated Harvard Business School graduates who succeeded when compared with those who did not perform as well after graduation. She spent a year working directly with students at the Harvard School of Business. At the end of the year, she recognized the key traits of mindfulness and mindlessness as a means for determining successful leadership behavior:

"Mindlessness is the application of yesterday's business solutions to today's problems.
Mindfulness is the attunement of today's demands to avoid tomorrow's difficulties."

Langer's definitions describing mindless and mindful leadership behaviors provide a new perspective regarding the application of lessons learned. Just because a strategy proved successful in the past does not mean that it will deliver the exact same results when applied again. The Harvard Business School students who were able to recognize the differences which existed in the current scenario as compared with the original situation were able to mindfully adapt the strategy to serve current requirements and achieve successful results. The students who applied the previous lessons learned to the current scenario without considering shifting circumstances were not as successful as their more mindful peers. Mindless leadership behaviors will not necessarily be called out as wrong since the managers and leaders are following organizational traditions. However, those people who can critically assess the situation and adapt solutions to address situational changes will be more effective leaders over the long run (Tippett 2015).

CONSIDER AND DELIBERATE

What are your greatest fears?

What can you do to alleviate this fear(s) or mitigate possible consequences?

When you are asked a question for which you have no answer, what is your initial response?

Do you behave differently if the question is asked by:

Your boss?

Your colleague?

Your reporting staff?

Your friend?

Reflect back on a complex work situation for which the efforts of many people integrated to produce the end product or service. Did you understand the processes that occurred prior to and after your work tasks? What were the reactions of people when you asked questions regarding their work tasks?

Describe a professional/personal effort that required you to adapt your original action plan because of a change in the environment or requirements.

The awareness of leaders regarding what they do not know enables them to reflect on previous experiences and adapt lessons learned to the current challenge to be ready to resolve uncertainties. When I find myself in a difficult position, I like to ruminate on a quote attributed to Marcel Proust, "The real voyage of discovery consists not in seeking new landscapes, but in having new eyes." Leadership is not necessarily about previous performance; it is about having the courage to seek out new solutions that are best suited to evolving environments.

Emotional Intelligence and Leaders

During the mid-1990s, Daniel Goleman introduced the concept of EI as a leadership tool. Initially, I was just as resistant as any self-proclaimed techy could be. I could not fathom a workplace driven by feelings. With the onset of the twenty-first century, I saw a shift in leadership theories, not necessarily practices, toward a more humanistic approach within the workplace. Technology stopped being the single-pronged business solution. Leaders were starting to be measured and accorded respect based on their technical competencies, cognitive abilities, and EI.

I decided it was time to take a second look at the theories and research supporting EI and learned what a powerful framework it provides to managers and leaders. Essentially, EI is a procedural paradigm that describes how people view and react to the world around them. Rita Carter, a science writer and broadcaster, describes the importance of this tool set, "Emotions are not feelings at all, but a set of body-rooted survival mechanisms that have evolved to turn us away from danger and propel us forward to things that may have benefit" (Kaku 2014, p. 34). There is scientific research validating those gut reactions, we have all tried to ignore over the years. In a nutshell, EI is important for leadership learning because it measures how leaders handle themselves and their relationships. It measures how leaders are perceived by themselves and others, the accuracy of their communications, and the manner they manage their emotions.

There is a Chinese parable in which a man goes to a spiritual teacher to discuss how to be happier in his marriage. The wise one listened to the man's complaints and told him, "You must learn to listen to your wife." The man took the advice to heart and returned after a month to say he had learned to listen to every word his wife said. He explained their relationship was improving, but he thought it could be better. The wise one smiled and told him, "Now go home and listen to every word she isn't saying."

Many organizational frameworks are similar to this man's home and in that there is always some level of underlying communications that is present, but left unsaid. Too many misunderstandings can happen because people only pay attention to information that is verbalized and not what people choose not to verbalize.

Work environments made up of isolated systems with minimal intraorganizational communications often evolve into disorder and declining performance.

When management does not address the practices that evolve from a dysfunctional work environment, people will become less and less engaged with constructive behaviors. These negative environments are the antithesis of a healthy, emotionally intelligent workplace.

The most effective visual image I can imagine to describe a dissonant leadership situation are the Dementors from the *Harry Potter* novels. For those not familiar with the J.K. Rowling series, it is set within a wizardry world that exists as a parallel environment within the late twentieth-century English society. The Dementors are dark, faceless creatures who guard the high-security wizard prison. Their function is to consume hope and happiness from the wizardry prisoners, leaving their victims in misery and despair. Leaders who encourage extreme competition so that people work against each other as opposed to working together to achieve their joint mission are Dementors in business attire. Like the Dementors described in the *Harry Potter* novels, dissonant leaders drain all hope and harmony from the workplace. Eventually the most talented people will find other employment and the spiral continues downward. The only means to banish these workplace Dementors is for emotionally intelligent people who are able to recognize what is not being said and who are able to convince people to redirect and propel them forward to a functional, productive working environment.

CONSIDER AND DELIBERATE

Who was the best manager you ever had? List three traits of this person which made you look forward to returning to work on Monday.

How have you modeled these traits to create a positive work environment?

Who was the worst manager you ever had? List three traits of this person that made you dread returning to work every Monday.

Given the wisdom of time and the opportunity for a *redo*, how might you behave differently to create a more constructive work environment?

Emotionally intelligent leaders are able to unify colleagues who may have been divided because of a blame culture by emphasizing a shared purpose, awareness of interdependence, and creating a relative narrative that will help guide them focus on what needs to be done.

EQ-i 2.0 Model—A Practical Approach for Emotional Intelligence

Reuven Bar-On led the psychometric research team that developed the EQ-i 2.0 model. He based his work on the premise that it is possible to identify and measure the emotional and social competencies that people demonstrate as they deal with daily activities. He believes that people who are emotionally and socially intelligent are able to cope with change by behaving in a realistic and flexible manner. These people demonstrate resiliency when facing challenges because in addition to being aware of their strengths and weaknesses, they have the ability to develop mutu ally supportive relationships with people who can help to achieve the desired goal (Bar-On 2017).

EQ-i 2.0	Composite	Sub-composite
	Self-perception	Self-regard Self-awareness Self-actualization
	Self-expression	Emotional expression Assertiveness Independence
	Interpersonal	Interpersonal relationships Empathy Social responsibility
	Stress management	Flexibility Stress tolerance Optimism
	Decision-making	Problem solving Reality testing Impulse control

Figure 7.1   EQ-i 2.0 composite matrix.

Bar-On identified five primary composites to measure an individual's emotional and social intelligence: (1) Self-Perception, (2) Self-Expression, (3) Interpersonal, (4) Decision-making, and (5) Stress Management. Figure 7.1 identifies these five composites and the associated subcomposites defined by the EQ-i 2.0 model that describe an individual's Emotional Quotient. The associated web-based feedback tool which is administered by Multi-Health Systems Inc. enables people to measure those intangible leadership skills that could not be previously measured.

The power behind the EQ-i 2.0 EI model is the dynamic interconnections between the five primary composites and the associated 15 subcomposites. A person will demonstrate various levels of engagement among the different composites. There is no optimal amount of EI; nor is it a static depiction of an individual's competency measure. A person's EI level will change on the basis of their level of maturity (aka wisdom) balanced against their personal and professional circumstances. The EI measurement is a reflection of a person's overall abilities at a given point in time. During periods of personal upheaval, someone may not maintain the same level of engagement in certain areas as they may have in the past, or again, in the future. Have you worked with someone who was going through a personal crisis such as an ill child or parent? Did you notice a difference in how they handled a typical situation while their priorities were divided?

The associated EQ-i 2.0 feedback report conveys an individual's engagement for each of the composite characteristics using a normal distribution scale (Bell curve) illustrated by Figure 7.2. The average population response is represented by the peak of the bell curve. People who have an engagement quotient that falls above the average for that characteristic will fall to the right of the curve. Those who engage less than the mean activity level fall to the left of the curve. The feedback

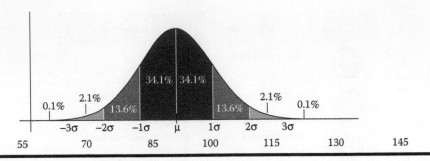

Figure 7.2   EQ-i 2.0 scores.

tool numeric scale does not represent a traditional grading system because there is no perfect score. In fact, an extremely high numeric score could mean that you are over-engaged and should consider adapting your behavior. For example, someone who measures in the third standard deviation to the right (light gray area) for Self-Awareness is so self-aware, they are not able to be aware of any one else. This extreme behavior is called narcissism.

Most people will demonstrate a variety of lesser and stronger engagement levels for the twenty different composite measures. Let me share two of my scores as an example. Impulse Control was one of the subcomposites that I scored an engagement level to the left of the mean. Reality Testing, on the other hand, was one of the subcomposites for which I demonstrated a high engagement level to the right of the mean. Translated into a practical scenario—although I might consider jumping off a bridge because it looks like an exciting experience, I will not because my well-honed analytical skills that are part of Reality Testing will stop me from acting on the impulse.

CONSIDER AND DELIBERATE

Of the fifteen subcomposites, identified in Figure 7.1, which four do you think are strengths for you?

Of the fifteen subcomposites, which two do you consider weaknesses for you?

Can you describe a situation when your identified strengths balanced out your identified weaknesses?

There is a reason one of the recurring themes of this book is the importance of recognizing your weaknesses as well as your strengths. A single trait does not drive a leader's success. It is the amalgamation of both positive and negative traits that shapes a person's leadership style. What I find enlightening about the EQ-i 2.0 feedback tool is that it provides people information about their engagement levels for the leadership traits that comprise the 5 composites and 15 subcomposites without judgment. We are who we are. Do I need to become more engaged with regard to Impulse Control? If I did, would I become less creative? It's my choice. If I had a different profession that required precise adherence to regulations, it might be in my best interests to explore different means to strengthen my Impulse Control engagement level.

Looking at a different example, introverts who may be less engaged with regard to Interpersonal Relationships may be highly engaged with regard to the Self-Expression composite because developing strong written communication skills will enable them to facilitate accurate and concise e-mail conversations. When an introvert is able to communicate effectively via e-mail, he/she is able to minimize the number of explanatory phone calls that are needed.

EQ-i 2.0 Composites and Leadership Practices

One of the reasons I am such a fan of EI is that I believe people can leave a training session and put these skills into immediate practice. The CEO of an IT start-up company came through the project management certificate curriculum I was teaching. A few months later, I called students to touch base and do a little research on any changes the curriculum might require. When he returned my call, he raved about the curriculum and the instructors. Positive comments are always welcome, but these meant a great deal because of the experience this gentleman brought with him to the training program. He told me the EI elective was his favorite class. When I asked why, he told me, "learning about stress management." I asked him to explain his comment. He shared he had difficulties with an employee who did not seem to respond to stress in what he considered an appropriate manner. He told me that the brief segment explaining how different people engaged with the Stress Management composite helped him to realize that his staff member recognized the urgency; however, she responded to stress differently than he did. Now, when there is a stressful situation, he is a better leader because he is able to understand her motivations and guide her appropriately so they can effectively work in tandem to resolve the problem. I find this story a perfect example of a person's ability to apply EI within the real world.

Table 7.2 explores a sampling of EQ-i 2.0 subcomposite traits by identifying how various levels of engagement translate into practical observations for leaders, to help them understand themselves and their team members (Stein and Book 2011).

Table 7.2   EQ-i 2.0 Engagement Levels

EQ-i 2.0 Subcomposite	Engagement Level	Practical Observations
Self-awareness		Recognition personal emotion drives certain consequences
	Medium-high	Respond logically because they recognize the trigger and potential outcomes
	Extreme high	Self-indulgent and makes decisions without regard to the situation reality
	Low	Often feel misunderstood and surprised by actions of others
Independence		Confident in the ideas you bring to a group
	Medium-high	Self-assured and autonomous thinkers

(Continued)

Table 7.2 (*Continued*) EQ-i 2.0 Engagement Levels

EQ-i 2.0 Subcomposite	Engagement Level	Practical Observations
	Extreme high	Renders authoritarian viewpoints and does not participate in group conversations
	Low	Indecisive and depends on others to make decisions for them
Interpersonal relationships		The ability to create mutual connections
	Medium-high	Strong relationship builders
	Extreme high	Unable to be alone and can inappropriately cross personal boundaries
	Low	Maintain a guarded persona with other people and prefer to be alone
Empathy		Awareness of and sensitivity for others needs
	Medium-high	Attuned to others feelings based on a given situation
	Extreme high	Overly involved in difficult situations, will withhold important information or avoid possible conflict to protect someone's feelings
	Low	Self-centered and emotionally detached from group
Flexibility		Ability to adapt thoughts and behaviors
	Medium-high	Recognize a change situation and are able to adapt appropriately
	Extreme high	Easily bored and unable to stick to a plan
	Low	Lack curiosity and are rigid in their resistance to change

CONSIDER AND DELIBERATE

Does anyone you know currently, or in the past, exhibit any of the extreme high behaviors described in Table 7.2?

If you find yourself on a new project in which the other team members seem to misunderstand your suggestions, what might you do to gain understanding and acceptance?

What might you say to someone who always seems surprised by the reactions colleagues have to the suggestions she/he makes?

Describe a recent situation in which you did not express yourself as effectively as you would have liked. What circumstances prevented you from communicating the information you wanted to share.

How might you facilitate a group conversation with people who do not want to participate in an open discussion because they either are convinced their answer is the only solution or simply want other people to make decisions without their input?

Describe a demanding or unexpected situation that arose within your work environment and the perceived consequences. What were your initial thoughts and physical sensations? As you worked through the stressful situation, how did your thinking and physical sensations adjust?

Describe a demanding or unexpected situation that arose as part of your personal life and the perceived consequences. What were your initial thoughts and physical sensations? As you worked through the stressful situation, how did your thinking and physical sensations adjust?

How does your ability to manage stress differ between work and home?

There are no simple answers for leaders who need to successfully navigate difficult situations. EI and the associated composites provide a framework to help you shape your perspective and consider how others may be shaping their perspective. During this reflection did you find yourself also considering aspects of the other attributes of Vulnerability, Educated Ignorance, and Mindfulness discussed within this chapter? These four leadership super powers leverage off one another to provide strong foundation to face off uncertainty and transform traditional thinking into new found strengths and innovation.

Integrating Leadership Super Powers

Along with the myriad of over-the-top expectations people have for their leaders, the concept of maintaining some sort of balance of well-being is perhaps one of the least tangible intangibles. Stewart Friedman, a professor at the Wharton School of Business, targets his research toward leadership and work/life integration. Friedman does not believe it is possible to maintain any sort of balance with regard to work or life because asking people to identify priorities by which to organize your time is unachievable. Instead he suggests that working toward a realistic level of integration among work, family, and social responsibilities will enable people to achieve success. Friedman's framework targets three parameters that encourage people to *Be Real, Be Whole, Be Innovative*. By focusing on these three goals, people are able to integrate the various aspects of their lives to create a complete persona. They are able to work toward being the best person they can be (Friedman 2014).

Working with complex matrix teams, organizational leaders must balance a variety of trade-offs to cope with conflicting priorities. Table 7.3 adapts Friedman's

Table 7.3 Channeling Leadership Super Powers

	Be Real	*Be Whole*	*Be Innovative*
Vulnerability	Know what matters Maintain authenticity	Envision mission Embody values	Challenge status quo Create cultures of innovation
Educated Ignorance	Align actions with values Clarify expectations	Build supportive networks Apply all resources	Convey values with stories Focus on results
Mindfulness	Listen actively to feedback	Weave disparate strands	Embrace change Envision new methods
Emotional Intelligence	Be accountable Resolve conflict	Manage boundaries Help others	Lead by example

suggestions for maintaining work/life integration identifying specific leadership methods that can be used to sort and channel the super power tools of Vulnerability, Educated Ignorance, Mindfulness, and EI via the mantra of *Be Real, Be Whole, Be Innovative.*

CONSIDER AND DELIBERATE

Select two of the methods from Table 7.3 that someone might use to channel one of the four leadership super powers.

Recount a time you witnessed or performed one of these behaviors as a means to guide a team effort that integrated individual activities to achieve a common goal.

Independent of whichever super power you choose to channel (Vulnerability, Educated Ignorance, Mindfulness, or EI), Table 7.3 provides a tool kit of leadership guidelines to help you:

1. Set priorities for yourself and your team
2. Select the best action for the current situation
3. Guide your solution finding efforts

Summary

Super Powers for Leaders

■ Recognizing fears and vulnerability enable leaders can use this knowledge as a catalyst to promote creativity, innovation, and learning.
■ Leaders are empowered and not inhibited by lack of knowledge. Educated Ignorance offers leader the opportunity to explore new methods and ideas.
■ Mindfulness is the attunement of today's demands to avoid tomorrow's difficulties. Just because a strategy proved successful in the past does not mean it will deliver the exact same results when applied again.
■ EI shapes perspectives describe how people view the world, measures how leaders handle stakeholder relationships, assures the accuracy of their communications, and authenticates how they manage their emotions.

Chapter 8

Conflict and Conversation

> Conversation empowers people to …
> Agreeability can cause …
> Complexity impacts conflict by …

Language can be a powerful tool for bringing people together. Collaborative conversations help people create innovative solutions by allowing them to build on each other's ideas. Earlier chapters have illustrated positive examples of people listening, learning, and expanding their view of the world. This chapter offers suggestions for how to use conversation to resolve conflict.

Pàdraig Ó Tuama, a theologian and poet, works with groups from around the world to resolve conflict using education, mediation, and dialog. He has helped people across the globe, including Europe, Africa, and Australia, to work through their hostilities using language to establish courtesy, inclusion, and generosity. Ó Tuama reinforces the importance of words when he shares, "… we infuse words with a sense of who we are. And so therefore, you're not just saying a word; you're communicating something that feels like your soul. And it might even be your soul. So the choice of a particular word is really, really important." This is why people need to recognize the limits of their ability to understand a situation and the need for them to be able to reach out to ask other people to help them to understand (Tippett 2017).

In his book, *In the Shelter*, Ó Tuama offers suggestions for creating calm within a troubled world. He included a favorite poem *Lost* by David Wagoner that contains the lines:

"Wherever you are is called Here,
And you must treat it as a powerful stranger,"

(Tippett 2017)

For a leader within any organization, this is a strong metaphor for handling the unplanned and unexpected that can cause rifts and conflict among colleagues and team members. How do you treat someone who moves into your neighborhood, who joins your faith community, who joins your sports team, who is newly hired by your organization? Do you immediately exclude them from the 'in crowd'? Or, do you welcome them and get to know them?

An executive vice-president of a large organization once told me, "Conflict is not bad. It is how you handle it that makes the difference." I was in my mid-20s when he shared that nugget of wisdom with me. My colleague and I reacted in an unprofessional manner during a meeting when we discovered one of his staff members had withheld information to discredit us. Visualize two young women in navy blue suits, stockings, and heels climbing across a conference table to throttle our false accuser. It was not a stellar moment. The two of us came to our senses before we were fully on top of the table. As my colleague and I were gathering our composure and offering apologies, Mr. D. (the Executive VP) looked us straight in the eye and said, "Ladies, ladies, conflict is not bad, it is how you handle it that makes the difference." In that instant, he was letting us know that he realized we had been 'set up' without blaming or embarrassing the person who did so. We quickly resolved the problem situation and the meeting continued according to the agenda without any more surprise attacks from our nemesis. Mr. D. could have played into the situation the staff member had tried to set up. He could have chastised us. He could have demanded we be removed from the account. Instead, he chose to mentor us and provided invaluable counsel that we could take forward.

Navigating conflict is never smooth. The ability to guide teams from conflict to a creative solution using the generosity of language and listening is another trait that separates leaders from managers. Pàdraig Ó Tuama established a retreat center named Corrymeela near Ballycastle in the west of Ireland. He tells a story about the name people originally thought meant *hill of harmony*. A few years after opening someone better versed in old Irish etymology interpreted Corrymeela as *place of lumpy crossings* (Tippett 2017). What better word describes the process of working through conflict than *lumpy crossings*? Conflict does not come with directions. Rarely is one conflict ever handled in the same manner as another. It takes an experienced leader to guide the team across the various *lumpy crossings* on the way to resolution.

CONSIDER AND DELIBERATE

Reflect on a recent conflict that took place within your professional or community environment. What action ignited the conflict?

Was the conflict a simple situation to resolve or was it a *lumpy crossing*?

Briefly describe *the wrong* that each party felt.

Describe the actual challenge that the eventual resolution resolved.

What actions were taken from the incident that initiated the resolution process to the end solution?

Sometimes, people postpone confrontation because they do not want to cause a scene or think the infraction is not worth making a fuss about. The challenge for leaders is to help their team to understand a courteous conversation that resolves a small problem from going a long way toward heading off an extreme conflict that may involve many people and take much time away from daily work activities.

Crucial Conversations

Kerry Patterson and his team of researchers developed the concept of crucial conversations as a means to maintain courteous dialog when you are approached to address a challenging situation immediately but are unprepared to do so. A crucial conversation usually occurs spontaneously during very public moments. An audience is always present to see your discomfort as you try to respond to the unexpected query. The topic is high risk because both of you feel very strongly about your opinions and the consequences of how you respond can have long-term effects.

Whether the situation is professional or personal, the timing and possibility of an explosive response is the foremost driver to determine how you respond. Have you ever had someone ask you a question about a project which wasn't going well, in front of the boss? Or, has someone angrily approached you when you were in the middle of a conversation in the office break room? People's default response to these types of queries either is to avoid conflict by saying nothing or to say or do something they might regret. Neither of these silent or violent reactions is a constructive means for handling conflict. The third option is to manage the situation with skill and grace.

Many times people do not realize their natural reactions until after the situation has occurred. The key to having a strong leader-like response at the ready is to practice your reaction to difficult situations before they happen. Table 8.1 below offers several common reactions to challenging situations. Read through each of the reactions to a challenging situation.

Table 8.1   Reactions to Challenging Situations Comparison Table

I Do This	Challenging Situation Reaction	Behavior
	I delay returning texts, e-mails or phone calls because I don't want to deal with the issue. If I ignore the person they will go away.	
	I am a very direct person and let people know when I think their ideas are stupid and will never succeed.	
	I try to counter harsh feedback with a positive comment or a self-deprecating joke.	
	I don't give harsh feedback when I don't want to deal with the consequences.	
	I tend to make absolute statements (e.g., everyone knows ..., it's obvious to anyone ...) when I believe my idea is a better way to handle a situation than any other one under discussion.	
	I try to make my point in an even tone, but sometimes the other person makes me feel so disrespected, I explode and say the wrong thing, which only makes me appear weaker.	
	I suddenly remember an urgent appointment when someone brings up a topic I do not want to discuss.	
	I find responding with sarcasm to ideas I don't think will work softens the blow to the other person.	
	I try to work through the conflict with others so we can work together to find out the real cause of the problem.	
	I will suggest a cooling-off break or reschedule the discussion when the people get caught up blaming others and stop looking for solutions.	

If you have responded using one of these methods place a check in the column labeled "I Do This." In the third column labeled "Behavior," identify the type of behavior as: avoidance, sarcasm, change topic, coercion, dismissive, or attack.

When you are confronted by a crucial situation, the only person you can control is yourself. The person initiating the query might present a sense of urgency; however, there is a no rule that states you must respond immediately to the query. A simple statement such as, "I appreciate your concern, let's block out some time tomorrow so that we can give the situation the full attention it deserves," defuses the situation, respects the person trying to bully your response, and offers you time to gather relevant information. Even if you are not able to postpone the crucial conversation, request a moment to gather your thoughts and focus on the essential priority by verifying your goals (Patterson et al. 2012, pp. 1–49)

1. For yourself
2. For your team, including any related stakeholders
3. For the relationship you share with the person confronting you

Stabilizing Crucial Moments

Strong leaders will not allow themselves to be bullied by an unexpected situation they are not immediately prepared to handle. They will stabilize an emotionally charged situation and arrange to schedule the discussion when all relevant information can be brought into the dialog. The secret to remaining calm during an emotionally charged moment is to remember it is the adrenaline, not necessarily the actual situation which is creating the charged stressful moments. Kerry Patterson and his research team studied several key performance indicators (KPIs) at several companies before and after their employees had gone through training and become skilled at crucial conversations. They discovered that significant measurable performance and financial results could be attributed to these new skills. A few examples of these measurable results enabled staff to (Patterson et al. 2012, pp. 111–114)

- Adjust budgets five times faster in response to financial downturns
- Save over $1500 and 8 hours of employee time for each crucial conversation engaged
- Increase effectiveness of virtual teams which is directly attributed to increases in trust and reduced time to complete work tasks

They discovered the official and unofficial leaders within these organizations who were able to build relationships, increase productivity, and demonstrate positive financial results were present in the moment. They stepped back from their ego and any self-serving emotions and focused on the immediate organizational needs to be met. They used simple negotiation and conversation tools to stabilize the potentially crucial moment.

The power of an *And* statement enables people to create a mutual goal while avoiding a negative situation. An example of this type of mutual goal is, "The project plan needs to be completed within sixty days and not exceed $100,000." As each party (e.g., sales and finance) participated in framing the goal, they share ownership and a mutual stake in the creative process to explore solutions to achieve success. A corollary to the mutual *And* is the contrasting *And* statement that juxtaposes two core points of view: "I don't want _____ to happen and I do want _____ to happen. An example of a contrasting *And* statement is "We need to figure out a solution that will enable us to finish the project by the end of month and continue staffing the telephone support lines."

Dr. Anthony Suchman made a career out of researching how to get conversations that are starting to spiral out of control back on track. His research identified two channels over which all work place conversations travel: the task channel and the relationship channel. He discovered that when task-focused disagreements spill over into the parallel relationship channel, emotions escalate and civil conversation constraints begin to break down. This breakdown is explained by neurological science. When a conversation appears to be headed toward disagreement, one or both parties activate their fear response. The hypothalamus, located in the front portion of the brain, triggers a fight-or-flight response which focuses attention toward survival needs. This involuntary fear response stops a person's capacity to think creatively.

On the contrary, alternative ideas brought forward for discussion can be erroneously interpreted as a personal attack because the person who offered the original idea perceives the feedback as a personal attack. Their primal response wants to know why you do not like them, because you do not like their idea. Dr. Suchman suggests a series of conversational statements which can navigate the conversation back toward separate task and relationship channels. These collaborative relationship-building value statements referenced by the acronym PEARLS are illustrated in Table 8.2 (Friedman 2016, pp. 24–25).

Table 8.2 PEARLS Collaborative Value Statements

Collaborative Value	Suggested Statement Sample
Partnership	I really want to work with you to solve this problem
Empathy	I hear your concern
Acknowledgment	I can see that you have put a lot of time and effort into this
Respect	I appreciate your expertise in this area
Legitimation	I agree, there are several aspects that should be considered …
Support	I want to help you successfully resolve this challenge

Driving off to Abilene

Sometimes in an effort to build an important relationship, people can be too agreeable and take action that no one actually wants to pursue. They agree because they think it was important to the other person with whom they wanted to build a relationship. Jerry B. Harvey, author of *The Abilene Paradox: The Management of Agreement*, classifies this behavior as the "inability to manage agreement" as there is no conflict present to manage. Harvey recognized this phenomenon of people taking action on what they believe others want as opposed to the actual preferences of the people involved while visiting his in-laws in West Texas as a newlywed during the early 1950s.

Imagine a hot dusty summer afternoon on a West Texas cattle ranch. The temperature is 104°F and there is no air conditioning. Harvey, his wife, and her parents were sitting on the screen porch lazily sipping cold drinks. Wanting to entertain his guests, his father-in-law announced they should take a 40 min drive to the nearest town, Abilene, for a steak dinner at his favorite restaurant. Harvey's wife responded, "Daddy what a wonderful idea!" Harvey could not imagine anything worse than a 40 min drive in a car with no air conditioning through a dust storm for a heavy West Texas steak dinner. He did not want to disagree with his wife and looked toward his mother-in-law to be the voice of reason and disagree. Mom simply agreed that it was too hot to cook. The group took the hot, dusty 40 min ride, ate a large dinner, and returned home driving through the hot, dusty evening. They returned to the screened porch, where his father-in-law complained loudly about how badly he felt after the heavy meal and the hot dusty drive home. His mother-in-law said, she only agreed to go because she thought everyone else wanted to go. Harvey's wife said she only agreed because she thought her daddy wanted to go so very much. His father-in-law said he only suggested it because he wanted to make sure his daughter and son-in-law enjoyed their visit. As it turned out, each of them would have been very happy whiling the evening away visiting on the screen porch and eating a light snack (Harvey 1988, pp. 17–43).

"Driving off to Abilene" is not the same as avoiding a crucial conversation in which people are fully aware of the other person's point of view, but don't want to deal with a potentially explosive situation. In an Abilene-like situation, the individuals assume a collective reality that is not true. In being too agreeable, the people involved look toward the easiest path forward and do not take the time to validate the data or notice nonverbal cues. When groups "Drive off to Abilene," their willingness to accept inaccurate information is a sure means that counterproductive results will be generated. Instead of building a strong relationship, the end result is that people become frustrated and often begin to blame each other as they try to identify the culprit who is responsible for the lack of communication and understanding.

CONSIDER AND DELIBERATE

Describe a time when you avoided a conflict situation. What were your reasons for not wanting to address the situation?

Describe a time when you were frustrated with how a colleague handled a difficult situation.

Looking forward, how might you use a crucial conversation technique to handle a similar situation differently?

Describe a time when you were with a group of people who made a decision to do something and later discovered no one actually wanted to participate, but went along because they thought everyone else wanted to participate.

How did you feel when you discovered no one wanted to participate, but went along because they thought everyone else wanted to pursue the activity?

Moving forward, what will you do when you find yourself in a situation in which you are going along with what you believe to be a majority preference?

When someone initiates a crucial conversation or a group starts off toward *Abilene*, the people involved in these conversations need to be able to speak freely. A strong leader is able to respond to a crucial conversation starter with an alternative choice both parties can evaluate, or a suggestion to schedule a time when all aspects of the decision can be properly reviewed. Either of these strategies will strengthen the ultimate solution and build collaborative relationships. The same applies to groups who are finding their way toward *Abilene*. Leaders need to help teams create safe environments in which they can discuss potential problems and share different opinions so that the decision becomes a group decision and not the desire of a single, strong-willed individual.

Creating Safe Conversation Spaces

Leaders may be working against established organizational norms when they begin to create an environment that encourages group members and stakeholders to participate in honest, open conversations that allow for hard topics to come to the forefront without setting off danger alarms. Organizational norms are established over the course of time based on a complicated matrix of actions and omissions.

These spoken and unspoken norms were never written as official policy; however, they were reinforced by the premise, "that is the way we do things here." Fear of potential repercussions inhibits people from questioning people who they believe are more experienced or possess greater authority. Add the complexity of multiple structures across a matrix organization, and it is easy to see how often the very best intentions can be misinterpreted.

Mark Kramer and Marc Pfitzer, work with governments and NGOs to help people create shared value to promote social change. They are familiar with the various patterns in which individuals trying to pursue a joint vision find themselves in a muddle because they misinterpreted each other's actions. Kramer and Pfitzer have successfully helped to coordinate the efforts of government, business, philanthropic organizations, and other affected populations to work together to coordinate their efforts and develop a plan that will have a successful collective impact to resolve a broad community challenge. Their techniques can be applied within a complex, matrix organization to create a safe shared space to pursue conversation. Successful collective impact efforts are based on the following five core elements:

■ Common agenda
■ Shared measurement system
■ Mutually reinforcing activities
■ Constant communication
■ Dedicated backbone support

The common agenda is described as a shared vision for change and joint approach that is collaboratively discussed and mutually agreed upon by all involved parties. Depending on the complexity of the populations involved, the initial phase to reach a common agenda can take several months to address the full spectrum of issues and policies. The shared measurement system should consist of a few Key Performance Indicators (KPI) by which people can measure their progress. The next element for success recognizes that each member of the collaborative group has a different set of skills and talents. The mutually reinforcing activities element recognizes that each person will contribute toward the greater whole based on their strengths, capabilities, and resources. It is very easy for a diverse group of people who are used to working within separate organizations to isolate themselves and assigned tasks from the rest of the collaborative group. Constant informal and structured communications are essential for people to maintain the synchronous rhythms that help to establish connections among coordinating organizations. The final essential element of dedicated backbone support is a separate independent staff that represents the interests of all organizational parties. The backbone is a group of people who are dedicated to the success of the initiative. They provide vision, supporting policies, shared measurement practices, and resource mobilization (Kramer 2016, pp. 81–89).

Once an organization has a shared vision and an endorsement to move forward, the leaders charged with launching change within the organization need to pay attention to how they can create a new system that will encourage new behaviors and reward improved performance effectiveness. Whether the organization is categorized as small-to-medium or enterprise conglomerate, leaders will juggle different types of demands as they try to work across a variety of departmental structures and management styles to help people adapt and grow to realize the new vision.

A leadership method that works within one type of organizational situation will not necessarily work across every situation. Leaders need to develop a variety of tools to respond appropriately to the situation at hand. David Snowden and Mary Boone developed a framework to recognize and appropriately address the complexities of an organization structure. The Cynefin (pronounced ku-*nev*-in) framework enables executives to rapidly assess, recognize, and respond to real world situations. Snowden and Boone have worked with a variety of complex organizations to validate the real word applicability of the Cynefin framework. These organizations include a broad range of private industries such as pharmaceutical companies and government agencies including the U.S. Defense Advanced Research Projects Agency.

The Cynefin framework is based on four types of organization environments: Simple, Complicated, Complex, and Chaotic. Table 8.3 illustrates this framework by identifying the characteristic traits for each environment and the role a leader should play within this type of environment. Common organizational danger signals are identified for each environment along with a suggested response to help resolve or mitigate the challenging situation. A fifth category labeled Disorder represents a jumbled conglomeration of all four environments. The authors suggest people address a Disorderly environment head-on by separating out structural entities and matching each structural element to whichever of the four basic organization environments best describes the situation (Snowden and Boone 2015, pp. 71–78).

A *Simple* environment is classified as an organization for which the executive management makes decisions based on facts derived from consistent operational patterns. The variables and the associated operational ranges are known entities. Leaders within these organizations focus on traditional management best practices and procedures. When staff begins to exhibit complacent attitudes, minimize operational problems, and hold onto rigid belief systems, leaders need to take action to reverse the organizational trend. The best response to jump start people who you feel are becoming too complacent and accepting of the status quo is to challenge their current beliefs and best practices. Play devil's advocate and push them to view the situation from another perspective and to reach beyond their comfort level.

A *Complicated* environment challenges executive management because every decision has several potential solutions. Organizational decisions are based on a series of analytical procedures that prioritizes risks associated with all of the possible consequences that could result from the selection of one of the solutions based on the known unknown variable. Leaders within a *Complicated* organization environment are dependent on advice from subject matter experts; however, they need to practice sound situational analysis to be able to logically assess conflicting expert opinions.

Table 8.3   The Cynefin Framework

Environment	Organization Characteristics	Leader's Role	Danger Signals	Needed Response
Simple	Fact-based management Consistent patterns Known knowns	Direct delegation Best practices Proper procedures	Complacency Oversimplification of problems Rigidity	Challenge beliefs Link into process Recognize best practice limits
Complicated	Diagnostics required Multiple possible answers Known unknowns	Situation analysis Expert advice Conflict assessment	Overconfidence Narrow focus Analysis paralysis	Challenge expert opinions Think outside box Involve stakeholders
Complex	Creative approach needed Unpredictability Unknown unknowns	Purposeful communications Idea generation Create safe spaces	Command and control Narrow focus Resolution urgency	Delegate solution tasks Encourage brainstorming Reflection
Chaotic	No cause effect Urgent decisions Unknowables	Order and alignment Solution seeking Communication	Tight controls Risk aversion No forward progress	Parallel teams Devil's advocate Separate tasks

Leaders need to watch out when recommendations are based on a very narrow focus and staff members become overconfident in their ability to resolve the situation. In these instances, leaders need to help staff members expand their perspective and thought process. Bringing knowledgeable stakeholders into these conversations may be an excellent means to challenge team experts to try alternate methodologies that might stimulate some innovative solutions.

A *Complex* environment challenges executive management because of the total unpredictability of the situation. Individuals who prefer traditional management styles will not survive within this environment because all decision variables are unknown. Leaders within this type of environment need to help employees develop a culture that supports a creative approach to solution analysis and decision-making. Leaders must render purposeful communications. If they are not careful, every casual remark will be taken to represent a change in direction. On the other hand, innovation is the foundation for success within *Complex* organizations and people look to leaders to establish safe places where staff members are encouraged to generate ideas. The danger signs for dysfunction within this environment are behaviors that inhibit creative conversations. Teams and individuals may take on a command & control demeanor which can limit the creative solutions. As people jockey for command & control positions, there is a tendency for every problem and its resolution to be an absolutely urgent priority. Every decision becomes an urgent fire drill. In this situation in which power games are commandeering the creative process, leaders need to step in and organize the team into smaller groups that will address different aspects of the solution. Group guidelines would encourage innovative solutions while defining the project scope and align the different group efforts when all of the elements are connected together. Managing urgencies can be a difficult challenge to balance. Once a leader realizes that there is not an immediate fire to put out, he or she can ask the team questions to encourage them to reflect on the challenge and realize in the calm after the initial storm a constructive solution and lessons that can be taken forward to avoid the situation in the future.

A *Chaotic* environment is just that—chaos. The executive management team is challenged because there is no logic to any decision within the organization, and they cannot depend on any type of precedent to help guide them. People need leaders to establish some level of order and help them to align their work activities with the organization's mission. Leaders within *Chaotic* environments are constantly troubleshooting as they try to establish a structure. Communication is a leader's key for success in these situations. It is human nature, when faced with chaos, for leaders to try to organize their little piece of the world; however, if leaders find their staff beginning to become risk averse, set tight controls, and have several excuses to explain why they are not meeting commitments, they need to take action. One suggestion to help leaders get their teams back on track in a *Chaotic* environment is to separate the work projects into achievable tasks and assign people to work in parallel teams with regularly scheduled meetings to update each groups progress and challenges.

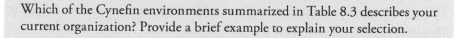

CONSIDER AND DELIBERATE

Which of the Cynefin environments summarized in Table 8.3 describes your current organization? Provide a brief example to explain your selection.

How do you or your organizational leadership align with the description of the leader's role within the organization? What other leadership traits do you recognize as being helpful within your organization at this time?

As an organization develops in size and function, there is a natural flow from a Simple to a Complicated environment. What types of professional development activity would you recommend for executives who are involved in the growth process?

Which of the four workplace environments identified in Table 8.3 do you think you would find most rewarding to work within? Why?

The Cynefin framework might dispel the adage about people rising to their level of incompetence. Not all leadership skills are transferable from one situation to another. However, a bit of professional development coupled with trial and error will enable leaders to expand their skill set and be successful across several types of organizational environments.

There is a fine line between guiding people and usurping the authority they thought they had been given. Open ended questions that encourage reflection and problem solving to drive innovation will empower your staff to accept your guidance and proactively explore innovative solutions. The following questions can help to guide rising leaders to develop their professional capabilities while they focus to achieve their desired goals (Beer et al. 2016, pp. 50–61).

■ Have the various team leaders established a coordinated strategy and vision?
■ Has the team collected feedback that reflects the actual organizational state and any barriers that are impeding the ability of people to achieve the stated goals?
■ What suggestions are being implemented to adapt organizational systems and practices to overcome the identified barriers?
■ Does the organization's culture and training programs support the new vision? If not, what adaptations are needed?

Accountability and Conversations

David Ignatius wrote an opinion article for *The Washington Post* which began with an anecdote about a former Central Intelligence Agency director visiting a top secret facility. During his visit, the staff gave him a t-shirt printed with the words, "Admit Nothing, Deny Everything, Make Counter Accusations" (Ignatius 2017, p. A17). The sentiment respectfully poked fun at some possibly well-earned stereotypes that various intelligence operatives may have earned over the years; however, how many of us have worked for managers who have done just that?

This is a good time to revisit the practice of trust among functional teams as described by Brené Brown which was discussed in Chapter 1. She describes functional teams as groups who respond to difficult situations by *turning toward each other* as opposed to *turning on each other* (Brown 2012, p. 101). Dysfunctional leaders encourage teams to blame each other when they model poor behaviors by refusing to admit they had a part in causing a problem much less take ownership to resolve it. Often these leaders will select and punish someone they want to be responsible. Accountability is demonstrated when a person takes responsibility for the cause and resolution of the problem. People do not need to be the designated leader to be accountable for a situation. However, being accountable for yourself and your team is a defining trait of strong leaders.

When Kerry Patterson and his research team were pursuing follow-up studies for their work with crucial conversations, they discovered people could learn how to address a difficult situation, but they could not always convince people to alter their behaviors. There will always be people who believe rules are meant for other people. Pursuing this line of inquiry, they discovered that 93% of the people they polled worked with a widely-recognized challenging-type of person, but people were afraid to address the problem for a variety of reasons. They were reticent to bring up behaviors which bent the rules or went against the norms because they feared the consequences. Patterson's team coined this phenomenon as an epidemic of silence. People were torn between what they thought were their only choices: say nothing and enable the poor behavior or potentially create a problem for themselves by calling out the bad behavior and facing negative consequences.

Broken commitments are one of the most significant behaviors people want to address when people discuss the need for accountability. Patterson's research team discovered that more than 70% of project managers were substantially behind schedule and over budget because no one spoke up when the executives created unattainable deadlines and budget forecasts. As the researchers queried the project managers being polled to discover how they found themselves in this dire situation, many of them replied that no one spoke up when the executive management established unattainable milestones. Those who did speak up were told there was no choice—make it happen. Rarely, did the executives who determined the schedule and budget accept responsibility when the milestones were missed. They also discovered when cross-functional teams were jointly working on a project, team members rarely would own up to problems that could affect the entire team. Colleagues working within a cross-functional group brought a difficult challenge to the project manager to discuss and resolve less than 20% of the time. This means more than 80% of the time, milestones were missed because no one communicated the information that project indicators were off-track (Patterson et al. 2013, pp. 1–14).

CONSIDER AND DELIBERATE

Describe a time when you did not own a mistake you made. What consequences caused you to defer your accountability?

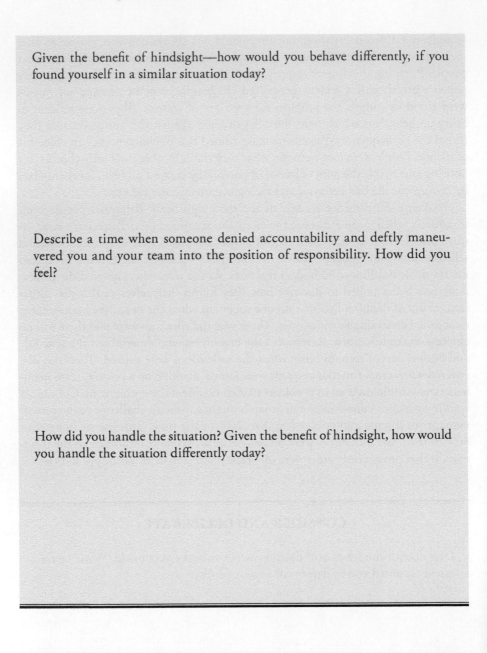

Given the benefit of hindsight—how would you behave differently, if you found yourself in a similar situation today?

Describe a time when someone denied accountability and deftly maneuvered you and your team into the position of responsibility. How did you feel?

How did you handle the situation? Given the benefit of hindsight, how would you handle the situation differently today?

It takes great courage to stand up and hold yourself and your team accountable for situations that seem outside the span of control. Sometimes it is easier to look the other way and dust up any evidence to the best of your ability. But what do you do when you are in a situation when paying attention or following protocol truly matters? How do you assure the people on your team and yourself take ownership and are answerable for mistakes as well as successes?

Traveling toward Accountability

Patterson and his team identified a set of *Crucial Accountability* procedures to help people learn how to guide a constructive conversation that encourages people to assume accountability. Similar to a crucial conversation, the first 30 seconds is essential. This brief moment will determine whether your conversation will be collaborative or combative. In fact, his team refers to this moment as the *hazardous half-minute*.

Before you begin the discussion, it is essential to have a plan. Identify and prioritize the behaviors you want to address. Review the potential consequences for each potential behavior. Then determine the behaviors you would like to occur from the conversation and the behaviors you do not want to continue as an outcome of the conversation. Finally, review what you do and don't want for yourself, others, and the relationship. The next series of steps vary slightly depending on the situation. They involve gathering facts, framing your story, and creating a conversation that will close the expectations gap (Patterson et al. 2013, pp. 149–198).

The Accountability Action model illustrated by Figure 8.1 expands on the four-step Path to Action model developed by Patterson's team. The first step is to assure you have gathered all of the available facts. Sometimes appearances can be deceiving. There is an adage, "each have less than one half of 100% of the truth," which means as hard as we try to gather facts and look through multiple lenses, we can still miss a key point to the story.

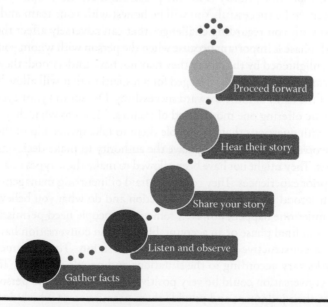

Figure 8.1   Accountability action model.

The next phase focuses on a leader's listening and observation skills. There is a strong chance that although you believe someone is not taking ownership for their actions, the person in question understands their role and the associated expectations differently. To their mind, they are performing well. There also is a chance that they are working through a personal crisis and have not told anyone about the external difficulties they are trying to juggle. Sometimes, people do not realize the consequences an action might have on the rest of the organization. Moreover on the contrary, there always is the chance that another person is dropping the ball and trying to frame someone else for their poor performance. Give people the benefit of the doubt, and once you have gathered the facts that identify poor performance and missed milestones, begin a conversation with the person involved in a safe space. Do not make this an inquisition. Begin with open-ended questions regarding the status of project. You can mention you have noticed they might have something on their mind. This is a conversation, continue to gather data by listening to what they are saying and not saying. Watch their body language.

Once you have an understanding of their perspective of the situation that concerns you, move on to the third phase. Explain that you perceive the situation differently. Use the facts you have gathered to back up the scenario you are shaping during this conversation. Maintain an even tone as you point out the gaps where your expectations do not align with theirs. It is helpful to write your concerns down prior to this conversation so that you will not forget an important point or lose your train of thought in case the person you are meeting with gets emotional or insists on defending their view of the role. The most important point to remember is to reassure them that your priority is for the project and all the people working with the project team to be successful. You will be honest with your team and hope they can be honest with you regarding challenges that can adversely affect the project.

The fourth phase is important because when the person with whom you are speaking becomes enlightened by the reality they may not have understood; they will begin to share information that can be leveraged for a resolution that will allow both of you and the rest of the team to move forward successfully. Do not roll your eyes when you perceive they are offering one more round of excuses. Listen to what they are sharing and respond with questions that will enable them to take ownership of the situation. Sometimes people do not realize they have the authority to make decisions or initiate certain actions. They might not have been allowed to make these types of judgments as part of their prior experiences. They could be afraid of incurring management's wrath. It might seem second nature to you to take action and do what you believe is needed to make the milestone happen; however, sometimes, people need permission.

The fifth and final phase of an accountability action conversation has the potential to make a constructive change in your organization. The agreement of what actions or tasks vary according to the situations and people involved. The summation of your conversation could be very positive as you provide the person feedback that you trust their judgment and provide them permission to which decisions they should make and which decisions should be brought to you. You might learn that certain aspects of the project were not going as planned and people were afraid to bring it to your attention. Or you might uncover the current job assignment is not a good fit and your conversation may turn to a discussion of next steps to help this person find a more successful path.

CONSIDER AND DELIBERATE

Reflect on a situation for which you did not have the appropriate experience. How did you balance your learning curve and accountability for the task completion?

Was the experience you described above successful? Was there a key success factor you can attribute to a positive experience? If the experience was not successful, in hindsight, how could you have handled the situation differently?

Describe a situation when you needed to help a new person gain expertise quickly in order for a task to complete. What was your role: a colleague, a supervisor, or a reporting team member?

Was the situation a positive or negative experience? If positive, what about the scenario made it successful? If it was negative, what were the key factors that caused a downfall? How might you handle the situation differently, if given a similar scenario today?

Given the benefit of hindsight—how would you behave differently if you found yourself in a similar situation today?

By encouraging continual, factual communication among everyone involved with a situation, strong leaders are able to monitor project status and requirements to assure people have the necessary skills and resources are available to achieve schedule, budget, and quality goals. Being accountable does not mean being all-knowing. Accountability is about addressing issues as they arise and working with your team to determine the most appropriate solution for the challenge at hand. Conflict will occur because people will bring different perspectives and priorities; however maintaining constructive conversation techniques with a focus on shared goals will maximize the opportunity for innovation and productivity improvements.

Summary

Conflict and Conversation

- Language is a powerful tool for bringing people together. Collaborative conversations have the power to help people create innovative solutions as they build on each other's ideas.
- Crucial conversations are high-risk situations in which the purpose of the instigator is to catch you off guard to either purposely make you look foolish publically or elicit an automatic yes answer to their request.
- Strong leaders stabilize crucial moments by remaining calm and using stabilizing techniques to diffuse the emotionally charged moment and create a safe space to constructively discuss the issue.
- *Driving off to Abilene* describes a situation in which people are unable to work toward an agreement because no one actually knows anyone else's preference and everyone involved is making decisions based on assumptions and not facts.
- Safe conversational spaces enable people to explore new ideas without fear of repercussions.
- Accountability is one of the traits of successful teams because all members take responsibility for the cause and resolution of challenges that affect the group and the achievement of shared goals.

Chapter 9

Leaders Are Human Too

Leaders enable good power ethics by …
Alpha leaders can …
Reflective learning action cycles drive …

Leaders are the ones who people count on to make things happen. Being the person to whom everyone looks toward for that new idea or what to do when the going gets tough is difficult work—and can be daunting. The load can become heavy and unmanageable, if you try to carry everything solely on your shoulders. Leaders are human beings; they are not perfect, and they will make mistakes as they navigate unfamiliar territory. The key is to practice forgiving yourself, your role models, and the people who work with you for mistakes and perceived weaknesses that might cause challenges along the way.

Carlo Rovelli, a quantum gravity physicist, aptly describes this quest for achievement in the preface of his book, *Reality Is Not What It Seems: The Journey to Quantum Gravity*. He writes, "This is not a book about certainties; it is a book about the adventure of moving toward the unknown" (Rovelli 2017). This is what true leadership is about; it is not about following well-defined rules. Leadership is about helping people navigate toward the unknown. Leadership goes beyond technology and buzz words. At the end of the day, leadership is defined by the people who travel the journey together. It is about maintaining alignment among the various organizational parts to achieve the shared mission and vision as a whole. It is not about taking the path less traveled as the poet Robert Frost suggests. It is about forging new paths by leveraging the strengths of your team to build the necessary roads and bridges.

During an interview, Rovelli described physics as a means of understanding the world, "I think we shouldn't reduce it to things. We should reduce it to a happening, and the happenings are always between different systems, always relations. Or always, like a kiss, which is something that happens between two persons" (Tippett 2017). Similar to physics, leadership is not about a person or a single group. Leaders need to be aware of the interconnections between organizational systems and the relationships that connect people to make success happen.

Paradoxical Behaviors

Dacher Keltner, a professor of psychology at the University of California at Berkeley, has researched executive behavior patterns over a twenty-year period to discover how people make success happen. He uncovered a troubling pattern that appeared in almost every study that he named *The Power Paradox*. This paradox describes a phenomenon that appears in a large number of executives who rise through formal leadership ranks, practicing leadership 'superpowers' and demonstrating traits such as empathy, collaboration, transparency, humility, and sharing. Once these leaders attained a position of power, they began to enjoy the accompanying privilege, and the qualities that helped them to rise through the ranks began to wane. People, who only a few months earlier had exhibited positive leadership traits, were now described as boorish. People, who once placed the needs of others before their own, were seen endorsing selfish and unethical actions.

Keltner's research team has run a variety of studies exploring elitist behaviors. One compared car owner behaviors based on the price of the car. They discovered that people who drove lower priced car models always yielded the right of way to other cars, whereas people driving luxury car models ceded the right of a way to another car just over 50% of the time. A global survey of organizations across 27 countries discovered that employees who were paid within the higher income tiers were more likely to consider bribes and other unethical behaviors to be normal business behaviors. Even more interesting is the impact that rude and selfish behaviors have on organizational productivity. Half of the people who responded to a poll, with 800 manager participants across 17 industries, admitted to purposefully decrease their work effort as a response to being treated rudely by a higher ranking organizational official.

Once Keltner realized the pervasiveness of the power paradox, he began to search for a means by which executives could avoid succumbing to these privileged behaviors. He discovered that people who recognized trigger warnings (e.g., frustration) were able to address their feelings in a logical manner. These executives were less likely to behave rudely and more likely to exhibit constructive behaviors. Keltner developed a series of practices he describes as *the ethics of good power* to help executives promote positive behaviors. The three practices of empathy, gratitude,

Table 9.1 Good Power Ethics

Good Power Ethic	Suggested Practices
Empathy	• Listen with your body turned toward and focused on speaker • Ask questions and paraphrase the answer you hear to confirm your understanding • Before meetings, take a moment to review what is happening in the life of the person with whom you will meet
Gratitude	• Publicly thank people for their efforts and contributions • Send colleagues e-mails or traditional thank you notes to appreciate a job well done • Celebrate small successes as well as the really big wins
Generosity	• Share the limelight and recognize everyone who contributed • Delegate important and high-profile responsibilities to staff • Seek opportunities to spend one-on-one time with people on your team

and generosity are a means to help leaders expand their ability to engage in simple relationship-building behaviors. Table 9.1 illustrates several suggestions to help practice the ethics of good power (Keltner 2016, pp. 112–115).

People notice everything about a leader's verbal, written, and nonverbal communications. In preparing this book, I asked many people to describe leaders who had made an impression on them. A friend who works for a global technology company described her current regional vice president. He visits the local office once every 4–8 weeks and makes a practice of walking around and saying good morning to each person in the office with appropriate small talk or memory from a previous visit. He might spend the rest of the day in a series of meetings with minimal breaks; but, he always takes 15–20 min to meet and greet the local employees before he begins his day. Previous regional vice-presidents have not extended this type of simple gesture to the staff. Their habit was to come into the office and get right to work. In our discussion, she believed the previous regional vice-presidents were fine managers who were focused on doing their jobs well; however, when she was asked to describe a leader, she did not think about the manager who drove the highest numbers, she thought of the person who noticed and appreciated the people who worked within his organization.

CONSIDER AND DELIBERATE

Reflect on a time when you were involved in a power paradox situation. Were you the person who received the privileged reward, or did you observe someone else assume a privilege as described by the power paradox?

Briefly describe the situation and include any thoughts or feelings you remember from the event. Would you behave differently today?

Describe the last time you practiced *empathy* during a professional or volunteer group activity. How did you feel at the time?

Describe the last time you practiced *gratitude* during a professional or volunteer group activity. How did you feel at the time?

Describe the last time you practiced *generosity* during a professional or volunteer group activity. How did you feel at the time?

Good power practices are excellent motivational habits to help one build relationships within your organization. The 15–20 minutes the aforementioned regional vice-president spent informally visiting with the staff at the field office was an effective team-building investment. In addition to letting the local staff know they are valued, the corporate officer was able to gain a snapshot of the local office culture. If there is an unsettling trend that requires higher level attention, the regional vice-president has established a baseline of the office culture and trust of the local staff should he need to make specific queries.

Alpha Personality Traps

Traditionally, executive positions were held by people who exhibited excessively ambitious and competitive Alpha or A-type personalities. Aspiring executives were encouraged to develop aggressive, tightly controlled management approaches that are often associated with Alpha-type leaders. Beta or B-Type leaders are not as aggressive as Alpha-type personalities and are noted for their collaborative work-styles. Kate Ludeman and Eddie Erlandson, authors of the *Alpha Male Syndrome* (2006), determined through their research that 70% of all senior executives could be classified as successful, Alpha-types. The executive profile developed by the research reflected people who enjoyed the stress and responsibilities of being in charge to making the tough decisions.

As Ludeman and Erlandson explored the various characteristics of Alpha-type executive behaviors, they learned the traits that have long been considered strengths for organizational leaders also could prove hazardous for organization growth within fast, changing environments. For example, Alpha-type executives are described as being able to act decisively and have strong intuition that allows them to make quick decisions. The flip side of the traits that enable Alpha-types to be self-confident and opinionated also can cause them to be close-minded, domineering, and intimidating

Table 9.2 Alpha Value/Risk Matrix

Alpha Attribute	Value	Risk
Self-confident/ Opinionated	Acts decisively	Closed minded and intimidating
Highly intelligent	Sees beyond the obvious	Dismisses and demeans colleagues who disagree
Action oriented	Produces results	Impatient; resists process changes for improved results
High-performance expectation	Sets and achieves high goals	Constantly dissatisfied; fails to appreciate/ motivate others
Direct communication style	Moves people to action	Generates fear and CYA culture of compliance
Highly disciplined	Extraordinarily productive	Has unreasonable expectations; misses signs of burnout
Unemotional	Laser focused and objective	Is difficult to connect with; does not inspire teams

when they find themselves in a collaborative situation in which they are not the sole decision-maker. Alpha-types also are known for being action-oriented. On the one hand, this attribute enables them to drive results; however, this same attribute can cause them to appear impatient and resistant to process changes that are not immediately evident. Seven common Alpha-type attributes are listed in Table 9.2 alongside the associated potential values and risks these behaviors bring to an organization.

Collaborative practices, which are innate to Beta-type leaders, are very difficult for Alpha-type leaders to perform. Ludeman's team has successfully coached many Alpha executives to become more effective leaders by helping them to become aware of their behaviors that could put the effectiveness and productivity of their organizations at risk. The basic premise is that in order for executive leaders to grow and change, they must take the time to become aware of their personal motivators, accept they will not have all of the answers, and explore new perspectives and contrary opinions.

Their core approach targets five goals for their clients to practice as they hone their Alpha attributes and learn how to successfully lead a collaborative twenty-first century workforce. The first goal is for the executive to admit vulnerability and begin to ask people for assistance. The second goal is to accept accountability for their impact on other people's performance. Type-A executives often do not recognize that their decisions can have an adverse impact on other people and departments within the organization. The third goal is to help the Type-A executive acknowledge and learn how to constructively manage their emotions. Learning how to balance positive and negative feedback is the fourth goal of the coaching process. Ludeman's team discovered that 80% of the feedback Alpha personality types deliver to their team focuses on negative measures. The purpose of this fourth goal is to help Type-A executives learn how to blend appreciative comments into their feedback flow so they can more effectively motivate and move their teams forward using a combination of critical and positive comments. The fifth and final goal of Ludeman's coaching process is to help executives recognize patterns associated with destructive behaviors. By helping an executive understand the root cause of their poor behaviors, he or she is able to let go of their defensive mode and become a leader who can help teams to focus on a constructive resolution (Ludeman and Erlandson 2016, pp. 62–72).

CONSIDER AND DELIBERATE

Do you work with an Alpha-type person in a professional or volunteer capacity? Briefly describe a challenge that you find difficult to manage when working together with them on a task.

How might you encourage them to recognize and address one of Ludeman's five goals (vulnerability, accountability, manage emotions, balanced feedback, and pattern recognition) to help Alpha-type people become stronger collaborative leaders?

Both Alpha and Beta personality type executives can become strong leaders. The type of person who is best to lead a project/task will depend on the specific situation. Although Alpha leaders need to practice delegating and working collaboratively, Beta personality types often can get caught up in the discussions and need to be coached to draw boundaries. Dr. Dana Ardi, author of *The Fall of the Alphas: The New Beta Way to Connect, Collaborate, Influence, and Lead,* describes several Beta-type leaders and the collaborative relationships they maintain with their second-in-command. One of the most famous of this type of partnership is between Warren Buffett, Berkshire Hathaway CEO, and Charlie Munger, Berkshire Hathaway Vice Chairman. Berkshire Hathaway is an American multinational conglomerate holding company headquartered in Omaha Nebraska. Buffett and Munger have worked together since 1982 and describe their roles within the company as costrategists, kindred spirits, and intellectual sparring partners who have developed a very efficient learning machine between them (Ardi 2013, pp. 188–191).

Military Generals and Mindfulness

Prior to reading *The Generals: American Military Command from World War II to Today* by Thomas Ricks, I believed the stereotype that military officers were all extreme Alpha-type leaders. I purchased the book so that I could better relate to the growing number of people transitioning out of the military who were attending my project management classes. Even though I can say "some of my best friends" served in the armed forces, I only had a superficial understanding of military organizations. I hoped that Ricks' book would provide me a better understanding of military strategy, the combat experience, and the impact of technology. Yes, I learned a great deal about those topics; however, my most interesting take away was an entirely new paradigm of what it takes to be a successful leader, one that can be applied in any type of organization: military, corporate, not-for-profit or educational.

I discovered that the generals who led U.S. troops through the four major conflicts of the latter half of the twentieth century were human beings who demonstrated intangible strengths and worked through a series of vulnerabilities. Some people might reference their actions as wise, mindful, or emotionally intelligent. I found their stories inspirational. The basic lessons I took from the generals who led the four major conflicts can be summarized as (1) Offer Second Chances, (2) Be Present, (3) Attune and Adapt, and (4) Enlarge Perspective. These leadership practices bring to mind the five EQ-i 2.0 emotional intelligence composites described in Chapter 7: self-awareness, self-expression, interpersonal, decision-making, and stress management. Figure 9.1 illustrates the parallel relationships between the EQ-i 2.0 composites and the mindfulness exhibited by the generals who led the U.S. Armed Forces during World War II, the Korean conflict, the Vietnam War, and the first Gulf War. The following sections expand my reflections from Ricks' excellent history.

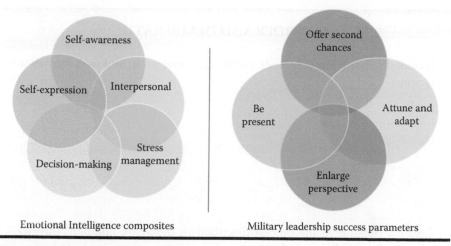

Figure 9.1 **Nontangible leadership skill comparisons.**

Second Chances

Offering second chances is the leadership lesson I pulled from Ricks' coverage of World War II and General George C. Marshall. In today's 24 × 7, fast-paced business world, how many people do you know who received a second chance after making a mistake? Marshall was a believer in training, education, and experience. Although he was known to remove inept officers from their command, there were instances during the war when Marshall realized that a previously high-performing general was becoming battle weary and initiated what some people reference as Marshall's 4R policy: relief, reassignment, recharge, return.

My observation of second chances is based on the 4R policy that Marshall demonstrated by recognizing that an officer under his command needed relief. He pulled them from active combat command and reassigned them to oversee a training camp for a year or two. The generals were competent; they simply needed a break from the stress of frontline combat. They were able to recharge and share their experience with the new recruits and rising officers. When they were ready, he brought them back to the battlefield where, in most instances, the returning general was very successful. Marshall rarely gave a third chance; however, his *err and learn* policy demonstrated the respect and trust he had for the people who reported to him (Ricks 2012, pp. 24–39; p. 451). Peter Dawkins, a West Point graduate, wrote an article for *Infantry* magazine in 1965 recalling the Marshall system and the military leadership framework it supported throughout the 1950s and the early 1960s. He wrote, "There was a time when an individual wasn't considered a very attractive candidate for promotion unless he had one or two scars on his record … If [a man] is to pursue a bold and vigorous path rather than one of conformity and acquiescence, he will sometimes err" (Ricks 2012, p. 213).

CONSIDER AND DELIBERATE

Describe a time when you made a mistake, corrected it, and took the lesson forward.

Create a timeline of the process described in the above question. Identify what you were feeling at each milestone (e.g., project start, mistake identification, confession, asking for help, problem resolution, next steps).

How did your relationship with the people involved change? Or, did your relationships remain the same?

Describe a professional or personal situation in which you were able to provide someone else a second chance.

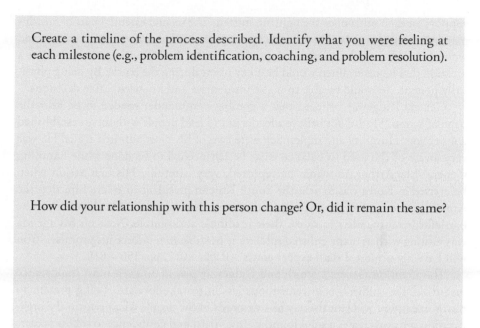

Create a timeline of the process described. Identify what you were feeling at each milestone (e.g., problem identification, coaching, and problem resolution).

How did your relationship with this person change? Or, did it remain the same?

Strong bonds develop between people who have worked through tough challenges together. Even civilians can feel as if they are in the middle of a fire fight when a deadline must be met and people realize the requirements changed, the budget did not, and several unforeseen challenges arose. Working together with someone to conquer what appears to be insurmountable makes the journey just a bit easier.

Being Present

The importance of being present is the lesson I took from the Korean War. General Douglas MacArthur tried commanding troops during the Korean Conflict from Tokyo. This was not necessarily a successful strategy, and there were many difficulties among the ground troops. General O.P. Smith who was the Marine commander in Korea and General Matthew Ridgeway who reported directly to General MacArthur took a more hands-on approach to rectify the poor troop morale, instill confidence, and improve combat leadership strategies.

Both Smith and Ridgeway believed in leadership by example. General Smith is credited with turning the tide of the Chosin Reservoir Campaign. Military historians attribute the success to his attention to details during the planning phase that preceded the attack. When he arrived in Korea, the troops were scattered across the geography. He consolidated the regiment so the Marines could provide better support for each other once the fighting started. The second essential part of Smith's plan was his order to build two air strips that enabled supplies to be flown in and wounded soldiers to be moved out quickly. Finally, Smith positioned himself at a location that he determined would be a key point during the battle. By being physically present, he would be able to make immediate and knowledgeable decisions.

General Ridgeway believed that a combat commander needed to be near the zone of action. He did not believe a leader could lead people without an established connection. Humility and diplomacy were two of his most admired traits. He was very aware of the need to balance what he felt needed to be done while handling General MacArthur, to whom he reported, very carefully. His first action when he arrived in Korea was to visit the South Korean president to assure him that the Americans were going to stay to help the South Koreans fight. He then visited the battlefield commanders to assess their readiness and morale. Next on his agenda was visiting with as many enlisted soldiers as possible to reinforce his promise, "You will have my utmost. I shall expect yours" (Ricks 2012, pp. 150–181).

The attention Generals Smith and Ridgeway placed on their most importance resource—the soldier—is a tremendous lesson to take forward. Being present to recognize and provide for the day-to-day needs of the people who perform the organization mission creates a solid foundation of trust and confidence to follow leaders as they maneuver forward during times of change and uncertainty.

CONSIDER AND DELIBERATE

Describe a time when you worked for a manager who 'phoned it in.' How did it affect the quality of your effort and that of your colleagues?

Describe a time when someone you respected asked you to put in some extra hours to achieve a goal. How did you feel when the request was made?

If you put in the extra time requested, how did you feel while doing the work? How did you feel when the task was complete?

General Ridgeway explained his aversion to standing on a platform to address troops, "I always disliked standing above people. I'm no better than they are. In rank, yes; but not as a man … In reviewing the troops … I always stood on the field, six to eight feet from the right flank of the unit going by. Then I could look into the eyes of the men going by. Looking into their eyes tells you something – and it tells them something, too" (Ricks 2012, p. 179). Being present is a two-way communication. Often no words are exchanged; however, the weight of the moment is not placed on a single person's shoulders. The exchange that happens when leaders are *present* lightens the load of the task owner's shoulders because they know they are not bearing the responsibility alone.

Attune and Adapt

The need to adjust strategies to meet the needs of the current situation is the basis for the *Attune and Adapt* lesson that I pulled from the Vietnam War. The early relationships between the U.S. military and the South Vietnamese were very difficult after the French military left Vietnam in the 1950s. There were some very difficult communication challenges. The early U.S. military leadership essentially ignored any discussions with the French transition team or the South Vietnamese. They were confident that the same tactics the U.S. military used to win World War II could be deployed with a successful outcome in Vietnam. With each change of U.S. military leadership, the World War II strategies and methods that proved successful in the open battlefields of Europe continued to be deployed. Even when General William Westmoreland and General William DePuy took their commands leading the Vietnamese campaign during the mid-1960s, they would not consider that an alternate strategy could be more effective. Robert Komer, a CIA officer assigned to the Vietnam pacification program in 1967, described the situation as, "… rather than shift to what it needed to do, the Army would continue doing what it knew how to do. … We went and fought the Vietnam War as if were fighting the Russians in the plains of central Europe for a very simple, straight-forward reason – that was what we were trained equipped, and configured to do" (Ricks 2012, pp. 217–264).

General Creighton Abrams, Jr. replaced General William Westmoreland in mid-1968. After consulting with a variety of stakeholders and initiating two-way discussions, he shifted the overall strategy to focus on pacification programs that emphasized protection of the villages and the Vietnamese people. He implemented a three-layer approach the marines had successfully deployed in remote areas. This strategy used large forces where necessary but deployed smaller guerilla-type patrols in the areas around population centers and deployed infantrymen in the villages to protect and build trust among the villagers. This new strategy surprised the Viet Cong who were not able to respond quickly and suffered significant setbacks (Ricks 2012, pp. 315–326).

The Vietnam War continues to hold a difficult place in American military history; however, the lessons to attune and adapt demonstrated by Abrams are invaluable at any time. A strategy that is successful within one environment will not necessarily work in another environment. Time, budget, resources, skill sets, and culture need to be considered as leaders select the best path to take. Making time to listen to stakeholder feedback and having the courage to view the situation from a variety of perspectives are basic tools that every leader should keep within easy reach.

CONSIDER AND DELIBERATE

Describe a time when you were involved in a project in which each person had a different view of what success looked like. Was the team successful or did they disband?

How did people handle their frustration?

How could the situation have been handled differently?

Once people have found a comfort zone that has been successful, it can be very difficult to explore any other means to accomplish a similar task. Their expertise and strengths were hard won, so why should they try a different way that might not be successful? Brand loyalty is a great example of this phenomenon. Do you continue to use a cleaning product because that is what your mother used? Are you hesitant to try a new food or a spice because it is different from what you are used to eating? It takes courage and tenacity to look for situational differences and alternate methods

Enlarge Perspective

Looking at a situation through multiple lenses to enlarge your perspective is the lesson I pulled from the first Gulf War. General Colin Powell and General Norman Schwarzkopf served as young officers in Vietnam where they learned the importance of becoming attuned to different circumstances and adjusting behaviors accordingly. In their effort not to repeat the mistakes made during the Vietnam War, they focused on coalition building and worked to involve stakeholders from different organizations and countries as they developed and adapted strategies appropriate for the Middle East.

One of the primary differences between the Vietnam and Gulf Wars is the diversity of cultures. From a cultural perspective, there were many similarities between the South and North Vietnamese peoples. Few, if any, of the allied and combatant countries involved with the Gulf War conflict shared a common culture, religion, or tribal affiliation. For both General Powell and General Schwarzkopf, success is attributed to their ability to look at a situation through the lenses of each stakeholder before deploying military action. There were few unforeseen consequences. If a mishap occurred, strong relationships had been built so that the appropriate leaders worked through whatever challenging issues arose. These collaborative actions are considered very positive demonstrations of modern leadership.

The network-centric technologies that were developing throughout the 1990s added another layer to complexity for combat zone decision-making. Schwarzkopf had to be careful not to allow his focus to become too narrow as he analyzed the large amounts of data available to him. If he allowed himself to deep dive into the vast amount of information, he might miss the forest for the trees. He also needed to view the data through several different lenses to assure his final decision considered the potential consequences from logistical, sociocultural, and political perspectives (Ricks 2012, pp. 367–394).

CONSIDER AND DELIBERATE

Reflect on a time when you had decided on a particular course of action but made changes because of new information that had not been available earlier. Briefly describe the initial activity.

How did you receive the significant information that caused the change of plan? Was it discovered as part of a correction? Or did it come from a source that was not included during the initial planning process?

What type of change process did you work through to incorporate the new information into the established design/plan? How confident did you feel when you made the decision to shift from the original plan?

With the benefit of hindsight, was there anything you could have done during the initial planning phase to have discovered the information sooner and incorporated it into the original plan?

Coalition building and stakeholder engagement are essential first steps for leaders who are venturing into new endeavors that will require innovative thinking. Bringing a variety of perspectives together will help you to broaden your perspective and practice empathy as you work to fully understand the consequences that might affect another group of people. Working with a coalition of stakeholders also will help you to identify and mitigate situations that could possibly prevent the success of your mission/project. Coalition members also can help you to build relationships with people who might not respond positively to you without a stakeholder's friendly intervention.

Leaders and Learning

The leadership lessons modeled by the generals who led the U.S. Armed forces are summarized within Table 9.3. Although the four themes are action-oriented, the generals who demonstrated these traits pursued these actions as part of a continuous reflective learning/planning cycle. Very similar to the Project Management

Table 9.3   Mindful Military Leadership Lessons

Major Conflict	*Leadership Lessons Learned*
World War II	Offer second chances
The Korean conflict	Be present
The Vietnam War	Attune and adapt
The First Gulf War	Enlarge perspectives

Institute (PMI) Plan and Do matrix, this continuous learning cycle begins with a defined undertaking with a focus on directing the resources available to you and assuring the actions align with the organizational values. The second part of this learning/planning cycle involves the process of eliminating as many known challenges before they happen. Once the execution of the plan begins, leaders must help to mediate unexpected risks as they arrive and flexibly adapt to changes. There will be times when the end vision needs to be adjusted and stakeholder expectations managed.

Leaders often are described as people who are curious and interested in different ideas and new perspectives. With so many workplace demands that require leaders to constantly seek information, analyze risks, and execute outcomes, how do they have the energy to pursue additional areas of knowledge? The answer is the energy they expend exploring new ideas is actually an investment in their personal development. Learning is an intrinsic means for leaders to maintain a broad perspective of the greater environment. In addition to honing a leader's competitive expertise, learning is a way to take a break from the details and responsibilities of day-to-day decision-making and enable people to see beyond the proverbial weeds.

Overlaying Leadership Models

Coming full circle back to the Diamond Leadership Model described in Chapter 1, Figure 9.2 overlays the four leadership lessons demonstrated by the twentieth-century generals on top of the intangible facets of trust, confidence, humility, integrity, and empathy. When General Marshall offered a second chance to a general who needed

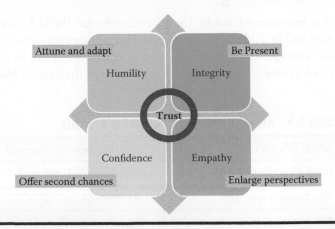

Figure 9.2 The diamond leadership model with military leadership overlay.

relief from the combat zone, he demonstrated self-confidence that enabled him to tell a direct report that he was confident in his abilities and offered him an alternative leadership assignment that allowed him to recharge. General Smith and General Ridgeway demonstrated integrity when they situated themselves in close proximity with their troops. Being present and paying attention to the day-to-day operational needs reinforced their commitment to their reporting officers and troops that they would be there to support them for the duration. General Abrams demonstrated humility as he practiced attune and adapt strategies when he established his three-level military strategy that was based on the unique qualities of the Vietnamese culture and geography. General Powell and General Schwarzkopf demonstrated empathy as they looked at the Gulf War challenges through multiple lenses. Before making a decision, they were able to consider how the ensuing consequences would affect each stakeholder group. Lastly, none of these generals would have been able to perform their duties as leaders, if they had not earned the trust of the officers who reported to them and the civilian stakeholders with whom they were affiliated.

Taking Leadership Lessons Forward

Personal growth is an essential element of the reflective learning/planning cycle. Referring back to the Organizational Clarity Model discussed in Chapter 2 (Figure 9.3), the relationship among entities, working groups, and leaders need to maintain a balance. As a leader and their working teams adapt to evolutionary changes, their perception of themselves and their relationships also might adjust.

Even without the impetus of a significant change, leaders should pursue some sort of personal reflection to verify they continue to be in balance with their

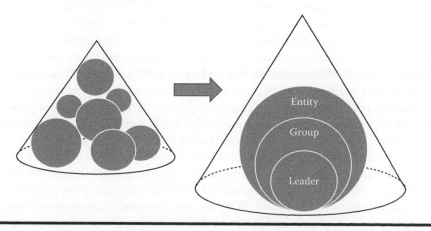

Figure 9.3   Organizational clarity model.

Figure 9.4   Ikigai life balance model.

organizations and associated working groups. Neil Pasricha, author of *The Happiness Equation*, wrote an article for *The Toronto Star* about why people should never retire. As part of his discussion, he referenced the fact that there is no Japanese word for retirement. This led to him discussing the term, Ikigai, that roughly translates "the reason you get out of bed in the morning" (Pasricha 2016). Ikigai is an ancient Japanese thought process that helps people to determine their life purpose.

It is based on four crucial questions that are meant to help a person realize how he or she perceives herself/himself and the world around them. These will evolve as a person matures and works through a variety of challenges and experiences. Figure 9.4 illustrates the interconnections of the Ikigai concepts that balance an individual's passion, mission, profession, and vocation (Winn 2014).

Pasricha cited a comprehensive study of 43,000 Japanese adults in his article for the *Toronto Star*. This study discovered that people who practiced Ikigai were happier, healthier, and had lower levels of stress than those who did not pursue this self-reflective practice. Pasricha's basic message is that finding an activity or cause that makes you feel purposeful and happy will help you maintain constructive connections with your surrounding environment. This applies to both your workplace and other personal pursuits. He recommends the long-proven Ikigai method as a means to help you find your personal north star aka your personal reason for being (Pasricha 2016).

As you work through this exercise to identify your passion and personal goals, remember this is a fluid process that may take several iterations. The end results are not always static. As your environment and worldview changes, your passions and personal goals may change as well. What was once extremely urgent to you may no longer be even a small priority. The four primary questions Ikigai practitioners ask themselves are

1. What do you love?
2. What are your talents?
3. What does the world (or someone) need?
4. What can you do to earn money?

Some people are able to wholeheartedly pursue their passion as their life mission. The Ikigai principles identify these people as being able to delight in life's fullness, but they will have no wealth. On the opposite end of the spectrum, people who place all of their emphasis on their professional vocations may have great wealth but will most likely maintain a nagging sense of emptiness. People who focus their energies on honing their talents will feel useful but may not take part within the greater society. Lastly, people who focus specifically on what the world needs will feel short-term excitement about their effort toward achieving the mission; however, they may have a sense of uncertainty regarding their longer term path. He/She is a very fortunate person who can lead a life in which these four elements intersect (Pasricha 2016). This intersection often is referenced to as a point of joy. So the truly big conundrum to work through is the answer to the question, what brings you joy?

CONSIDER AND DELIBERATE

Given a single day without any responsibilities or places to be (yes, the children are with a trusted caregiver), what would you do?

Given one week with the same freedoms as the question above, and an unlimited budget, what would you do?

Who is the one person you have met within your lifetime, even briefly, with whom you would like to spend an uninterrupted afternoon talking? Why?

Select the 3–5 people you consider closest to you (friends, family, or colleagues). What one gift do they share with you? And, what is the one gift you share with each of them?

Everyone has their own definition of joy. Sometimes we get so busy with the workaday world and making sure the people who work with us are successful, we forget to prioritize our own personal development. Taking a few minutes of reflective time each day to practice Ikigai is one way for you to focus, albeit briefly, on your professional and personal desires and what is required to achieve them. These private reflective moments may enable you to uncover thoughts and possibilities that lurk in the recesses of your brain waiting for the time, place, and opportunity to be revealed.

As this final chapter concludes, please remember the following hard-learned lessons:

■ Have high standards and practice kindness to yourself and others
■ Be generous with your attentions
■ Be direct in your conversations—say what needs to be said
■ Different situations will require different leadership skills/strengths
■ Mistakes happen—forgive yourself and others
■ Acknowledge failure as a learning opportunity
■ Offer second chances, be present, attune and adapt, enlarge your perspective
■ Continue to seek out your passion

Summary

Leaders Are Human Too

- Leadership is not about a single person or group. It is about the inter-connections between organizational systems and the relationships that connect people to make success happen.
- A power paradox is a phenomenon that appears when an executive who has risen through the ranks practicing collaborative leadership traits begins to demonstrate behaviors based on their newly acquired rank that could be described as selfish and unethical.
- Leaders often are divided into categories based on their personalities and associated behaviors. Alpha-type leaders tend to be very ambitious, competitive, and lead from a top–down approach. Beta-type leaders tend to be less aggressive and encourage collaborative workstyles. Neither type is good or bad. Each can be more effective in different environments.
- The Military Leadership Success model is based on four parameters pulled from each of the four major conflicts of the twentieth century. These four parameters are—offer second chances, be present, attune and adapt, and enlarge your perspective.
- The most effective leaders are those people who never stop learning. Ikigai is an ancient Japanese process that offers people a structured methodology that can be applied to both your workplace and personal pursuits to help you determine the direction of your learning efforts.

Bibliography

Ardi, D. (2013). *The Fall of the Alphas: The New Beta Way to Connect, Collaborate, Influence – And Lead.* New York: St. Martin's Press.

Bar-On, R. What it means to be emotionally and socially intelligent according to the Bar-On model. Retrieved February 14, 2017, from http://www.reuvenbaron.org/wp/43-2/

Beer, M., Finnstrom, M., and Schrader, D. (2016, October). Why leadership training fails – And what to do about it. *Harvard Business Review.*

Brown, B. (2012). *Daring Greatly.* New York: Avery.

Cloud, H. (2013). *Boundaries for Leaders.* New York: Harper Business.

Covey, S. M. R. (2006). *The Speed of Trust.* New York: Free Press.

Dayrell, E. (1910). Folk stories from Southern Nigeria. Retrieved January 27, 2017, from http://www.sacred-texts.com/afr/fssn/fsn18.htm

Detert, J. R. and Burris, E. B. (2016, January–February). Can your employees really speak freely? *Harvard Business Review.*

Dixon, M., Ponomareff, L., Turner, S., and DeLisi, R. (2017, January–February). Kick-ass customer service. *Harvard Business Review.*

Friedman, R. (2016, Fall). Defusing an emotionally charged conversation with a colleague. *Harvard Business Review On Point.*

Friedman, S. (2014, September). Work+home+community+self. *Harvard Business Review.*

Garvin, D. A. and Margolis, J. D. (2015, January–February). The art of giving & receiving advice. *Harvard Business Review.*

Govindarajan, V. (2016, November). Planned opportunism. *Harvard Business Review.*

Grant, A. (2013). *Give and Take: Why Helping Others Drive our Success.* New York: Penguin Books.

Grant, A. (2016). *Originals.* New York: Penguin Books.

Hadfield, C. (2015). *An Astronauts Guide to Life on Earth* (pp. 105–106). London, UK: Pan Books (Original work published simultaneously in 2013 by Random House: Toronto, Canada and Little Brown: New York).

Halpern, J. M. (1957). Interview concerning Xieng Mieng. Retrieved February 2, 2017, from http://credo.library.umass.edu/view/full/mufs001-ls-fn-i079

Harford, T. (2016). *Messy: The Power of Disorder to Transform Our Lives.* New York: Riverhead Books.

Harvey, J. (1988, Summer). *Organization Dynamics.* New York: The American Management Association.

Heath, C. and Heath, G. (2010). *Switch* (pp. 1–100). New York: Broadway Books.

Heidari-Robinson, S. and Heywood, S. (2016, November). A practical guide to a misunderstood – And often mismanaged – Process. *Harvard Business Review*.

Ignatius, D. (2017, April 5). Trump's shell game on Russia. *The Washington Post*.

Johnson, S. (2002). *Emergence: The Connected Lives of Ants, Brains, Cities, and Software* (pp. 29–33). New York: Scribner.

Kaku, M. (2014). *The Future of the Mind*. New York: Doubleday.

Keltner, D. (2016, October). Managing yourself don't let power corrupt you. *Harvard Business Review*.

Kramer, M. and Pfitzer, M. (2016, October). The ecosystem of shared value. *The Harvard Business Review*.

Ludeman, K. and Erlandson, E. (2016, Fall). Coaching the alpha male. How to work with toxic colleagues. *Harvard Business Review On Point*.

Magids, S., Zorfas, A., and Leeman, D. (2015, November). The new science of customer emotions: A better way to drive growth & profitability. *Harvard Business Review*.

Marcus, B. (2004). *The Anchor Book of New American Short Stories* (p. xii). New York: Anchor Books.

Maxwell, J. C. (2014). *Good Leaders Ask Great Questions*. New York: Center Street.

McGregor, J. (2016, December 11). For Lego CEO, rebuilding was no children's game. *The Washington Post*.

Morgan, G. (2006). *Images of Organization*. Thousand Oaks, CA: Sage Publications.

Murphy, E. (1996). *The Genius of Sitting Bull: 12 Heroic Strategies for Today's Business Leaders*. Englewood Cliffs, NJ: Prentice-Hall.

Noah, T. (2016). *Born a Crime* (pp. 55–56). New York: Spiegel & Grau.

Pasricha, E. (2016). Why North Americans should consider dumping age-old retirement. The Star.com. Retrieved April 24, 2017, from https://www.thestar.com/life/relationships/2016/09/06/why-north-americans-should-consider-dumping-age-old-retirement-pasricha.html

Patterson, K., Grenny, J., McMillan, R., and Switzler, A. (2012). *Crucial Conversation*, 2nd ed. New York: McGraw-Hill.

Patterson, K., Grenny, J., McMillan, R., and Switzler, A. (2013). *Crucial Accountability*, 2nd ed. New York: McGraw-Hill.

Project Management Institute. (2013). *A Guide to the Project Management Body of Knowledge*, 5th ed. Newtown Square, PA: Project Management Institute.

Reeves, M., Levin, S., and Ueda, D. (2016, January–February). The biology of corporate survival. *Harvard Business Review*.

Ricks, T. (2012). *The Generals: American Military Command from World War II to Today*. New York: The Penguin Press.

Rovelli, C. (2017). *Reality Is Not What It Seems: The Journey to Quantum Gravity*. New York: Riverhead Books. Translated by S. Carnell and E. Segre.

Sacks, J. (2007). *The Home We Build Together – Recreating Society* (pp. 4–5). London, UK: Continuum.

Schein, E. (2004). *Organizational Culture and Leadership*, 3rd ed. San Francisco, CA: Jossey-Bass.

Sinek, S. (2009). *Start with Why*. New York: Penguin Books.

Sinek, S. (2014). *Leaders Eat Last* (p. 20). New York: Penguin.

Smith, W. K., Lewis, M. W., and Tushman, M. L. (2016, May). "Both/and" leadership. *Harvard Business Review*.

Snowden, D. J. and Boone, M. E. (2015, Winter). A leader's framework for decision making. *Harvard Business Review On Point.*

Stein, S. and Book, H. (2011). *The EQ Edge – Emotional Intelligence and Your Success.* Mississauga, Canada: Jossey-Bass.

Tippet, K. (2014a, January 23). Teilhard de Chardin's "planetary mind" and our spiritual evolution. *On Being.*

Tippett, K. (2014b, October 23). Seeing the underside and seeing god: Tattoos, tradition, and grace. *On Being.*

Tippett, K. (2015, September 10). Science of mindlessness and mindfulness. *On Being.*

Tippett, K. (2016, August 8). The universe participates in the mystery of god. *On Being.*

Tippett, K. (2017a, March 2). Belonging creates and undoes us both. *On Being.*

Tippett, K. (2017b, March 16). All reality is interaction. *On Being.*

Winn, M. (2014, May 14). What is your Ikigai? Retrieved May 1, 2016, from http://theviewinside.me/what-is-your-ikigai/

Wong, Z. (2013). *Personal Effectiveness in Project Management* (pp. 32–33). Newton Square, PA: Project Management Institute.

Zak, P. (2017, January–February). The neuroscience of trust. *Harvard Business Review.*

Index

Printed and bound by CPI Group (UK) Ltd, Croydon, CR0 4YY

17/10/2024

01775709-0001